W.J Sterland

The Birds of Sherwood Forest

With Notes on Their Habits, Nesting, Migrations, etc.

W.J Sterland

The Birds of Sherwood Forest
With Notes on Their Habits, Nesting, Migrations, etc.

ISBN/EAN: 9783337026011

Printed in Europe, USA, Canada, Australia, Japan

Cover: Foto ©berggeist007 / pixelio.de

More available books at **www.hansebooks.com**

THE BIRDS

OF

SHERWOOD FOREST.

WITH NOTES ON THEIR HABITS, NESTING,
MIGRATIONS, &c.

BEING A CONTRIBUTION TO THE NATURAL HISTORY OF THE COUNTY.

BY

W. J. STERLAND.

WTIH FOUR ILLUSTRATIONS BY THE AUTHOR.

LONDON:
L. REEVE & CO., 5, HENRIETTA STREET, COVENT GARDEN.
1869.

LONDON:
SAVILL, EDWARDS AND CO., PRINTERS, CHANDOS STREET,
COVENT GARDEN.

"Faunists, as you observe, are too apt to acquiesce in bare descriptions and a few synonyms. The reason is plain, because all that may be done at home in a man's study; but the investigation of the life and conversation of animals is a concern of much more trouble and difficulty, and is not to be attained but by the active and inquisitive, and by those that reside much in the country. Men that undertake only one district, are much more likely to advance natural knowledge, than those that grasp at more than they can possibly be acquainted with. Every kingdom, every province should have its own monographer."—GILBERT WHITE.

PREFACE.

THE substance of the following pages originally appeared in the well-known natural history columns of the *Field* newspaper, during the years 1865-6 and 7. They were fortunate in attracting considerable attention, and in eliciting from numerous readers a wish for their separate publication.

With some diffidence I now comply with that desire, for my little work makes no pretension to be an exhaustive history of the birds it treats upon, but is a simple record of the results of twenty years' observations in a district of great natural interest. My occupations took me much out of doors, and I omitted no opportunity of jotting down every fact that came under my notice, that might bear on the life-history of our feathered friends. Some of my notes are but bare records of the occurrence of a species; others I trust may be found to possess a fuller interest.

I have carefully revised the original papers, and have added much additional information, the result of later observations. The introductory chapter on the forest will, I hope, give my readers some idea of the district in which my labours have been carried on.

One word with regard to an objection which has been made to local histories. It is not pretended that the Birds of Sherwood Forest are peculiar to that district, but local naturalists are confined to local boundaries, though it is obvious that these must often be arbitrary, or even imaginary; yet it is only by close and continuous local observation that the presence of a species in a particular district is detected, and its range determined, and thus our knowledge of the birds of the whole country is made more complete. I wish every county had its ornithological biographer, for we should thus not only become better acquainted with the habits and manners of our feathered neighbours, but much light would be thrown on a point which is confessedly obscure—viz., their local and general migrations, and the causes by which they are influenced.

I cannot conclude without expressing my obligations

to Sir William Jardine. It was that veteran naturalist who kindly stimulated my early efforts, and aided me by his suggestions, and to our correspondence on natural history topics, carried on during many years, I owe much pleasure and profit.

W. J. S.

April, 1869.

CONTENTS.

	PAGE
PREFACE	vii
CHAP. I. SHERWOOD FOREST	1
II. BIRDS OF PREY	14
III. PERCHING BIRDS	46
IV. PERCHING BIRDS—*continued*	123
V. GAME BIRDS	180
VI. WADING BIRDS	187
VII. WATER BIRDS	204
APPENDIX	235
INDEX	241

ILLUSTRATIONS.

BLACK REDSTART	*Frontispiece*
INOSCULATED FIR	11
TREE SPARROW	103
HEN IN MALE PLUMAGE	183

THE BIRDS
OF
SHERWOOD FOREST.

CHAPTER I.

INTRODUCTORY.

THE ancient forest of Sherwood was, in days gone by, of far greater extent than it is at present, spreading northwards from Nottingham to beyond Worksop, a distance of from twenty-five to thirty miles, and ranging from seven to ten miles in breadth. Its capabilities for the chase were indeed so great as very early to attract the attention of royalty, and the estate surrounding the palace of Clipstone (which, indeed, having been a park before the Norman Conquest, is one of the oldest in England), was seized by William the Conqueror, and made a royal demesne. I am not aware whether he enlarged or rebuilt the residence of Clipstone, but he was partial to its seclusion when he relaxed from the cares of government, though now a few shapeless ruins alone stand to attest its former grandeur and importance.

It was here, on his return from the Crusades, that Richard Cœur de Lion received the congratulations of the King of Scotland, and it was a favourite resort of both John and Edward I. Edward II. kept his Christmas here in 1315-16, and Edward III. granted a charter to the town of Nottingham, from Clipstone, in the first year of his reign.

Three abbeys had their sites within the limits of the forest—viz., Welbeck, Newstead, and Rufford, and they were each selected by their founders with that exquisite taste which distinguished the "monks of old," though the holy fathers paid more regard to the requirements of the kitchen in their selection, than to the mere beauty of the scenery that surrounded their abodes.

The abbey of Welbeck was founded in the reign of Henry II. by Thomas de Cuckney for Premonstratensian canons, and the adjoining manor of Cuckney was settled upon it by John Hotham, Bishop of Ely, in 1329. It was the head of all the houses of this order in England and Wales.

Newstead was founded by Henry II. for a colony of Augustine monks in 1170, and it was granted at the dissolution to Sir John Byron, who converted it into a residence. This Sir John was an ancestor of the poet, who thus in Don Juan described the building:—

> "The mansion's self was vast and venerable,
>   With more of the monastic than has been
> Elsewhere preserved; the cloisters still were stable,
>   The cells too, and refectory I ween.
> An exquisite small chapel had been able,
>   Still unimpaired to decorate the scene:
> The rest had been reformed, replaced, or sunk,
> And spoke more of the baron than the monk."

Rufford Abbey was inhabited by Cistercian monks, for whom it was founded in 1148 by the Earl of Lincoln, and after the dissolution of monasteries it came by exchange into the hands of the Earl of Shrewsbury. In addition to these three abbeys, there was a priory of Augustins at Worksop, so that the forest was almost as well supplied with monks as with deer.

The northern half of the forest was more thickly wooded than the southern, and, indeed, is so now, that part around the town of Ollerton still retaining its ancient trees, and forming the most perfect specimen of a forest of the feudal times existing in England. Besides the extensive tracts still unenclosed, large portions have been reclaimed from their sylvan wildness, and form parks which, for extent and beauty, have few equals in the country. Thoresby, Clumber, Welbeck, Rufford, and Worksop Manor, all cluster round the present forest, of which in bygone years they formed no insignificant portion. These extensive domains, whilst retaining much of their original beauty, have been greatly improved by their owners, and their wild features have given place to a softer loveliness. Much of the open heathy ground has been covered with thriving woods and plantations, especially of oak, an enormous number of these trees having been planted by the late Duke of Portland, who took great delight in the work. Large sheets of water, swarming with wild fowl, ornament each park, and the game in all is closely preserved.

The whole of this district, in addition to its charm of great natural beauty, possesses an historical interest, which, in spite of these utilitarian times, is not soon likely to die. I have spoken of the numerous monks, who, loving good cheer, fixed themselves where it could be easily obtained, but I must not omit to mention one who was their sworn enemy. Robin Hood has won a name in our nation's annals, whilst his gallant esquire, Little John (a native of the county), with Will Scarlet,

Maid Marian, and the jovial Friar Tuck, are personages with whose doings in the glades of "Merrie Sherwood," ballad and song have made all familiar. It would require but little play of the fancy to bring them back to their former haunts, for a large portion of the forest is comparatively little changed from what it was in the days of the renowned freebooter; the same huge oaks, whose gnarled and rifted trunks bear witness to their antiquity, still lift their giant arms aloft in sturdy grandeur. Furze, and bracken, and heather, cover the ground, and with the young self-sown trees, form dense thickets, where the red deer might hide securely, and it needs but to add the ring of the bugle, with the twang of the bowstring, and the stalwart figures in Lincoln green, to complete the picture of the past.

Geologically, the greater part of the forest lies in the New Red Sandstone; the northern extremity, however, is included in the Magnesian Limestone, which, commencing at Worksop, runs by Welbeck to Warsop, near Mansfield, and then taking a more northerly course, it joins the coal measures at Radford, near Nottingham.

The whole of the district, from Ollerton to Worksop on the one side, and to near Mansfield on the other, is closely wooded, parts of this area being unenclosed and covered with aged oaks, whilst here and there woods of more recent growth, and still younger plantations, are interspersed. Underneath the trees is a dense growth of the Common Ling, or heather (*Erica vulgaris*), mingled with patches of the Cross-leaved Heath (*E. tetralix*), and large tracts of furze, while here and there the common Broom reminds us of the origin of the surname Plantagenet, *planta-genista.* Over the whole district the Bracken (*Pteris aquilina*) grows abundantly; on a few exposed parts it is comparatively dwarf and stunted, but it generally is exceedingly luxuriant, and so dense that many a band of men like the highlanders of Roderick Dhu might hide in its cover without suspicion. Some stalks of this fern I have gathered which measured eleven feet from the ground to the topmost frond.

The peculiar arrangement of the ligneous fibres of the bracken is worth noticing. These fibres are so disposed, that on making a section of the root an exact representation is obtained of an oak tree, and their colour being blackish brown, they stand out with great distinctness amidst the pale cellular tissue. Often as a boy have I delighted to pull the larger stalks that I might have a better representation of "King Charles in the Oak," as children love to call it, and as by varying the angle at which the stalk was cut, I could at pleasure make the tree dwarf or tall, it was always a source of amusement.

The old woods and the various parks I have mentioned which cluster around, are now the only vestiges of the ancient forest, and glorious remnants they are. The venerable oaks possess a beauty that is heightened instead of being defaced by decay, and carry back our thoughts to the scenes they have witnessed in the bygone history of our country. Undoubtedly many of these trees are of great antiquity, some of them, the giant patriarchs of the forest, are a thousand years old at the least. A discovery of a singular link between the past and the present was made in 1786, in cutting up some trees for the royal dockyards. In one, at a depth of twelve inches from the outside, the letter I was found, surmounted by a crown, with blunt radiated points, such as in old prints was represented on the head of King John. In another, about the same distance within the tree, the letters I. R., James Rex, were revealed, while in a third, three feet three inches from the centre, and about nine inches from the outside, were the letters W. M., and a crown, the initials most probably of William and Mary. I believe these relics are now preserved in the arsenal at Woolwich.

Some of the trees possess a reputation distinct from their compeers, and are worthy of separate mention. Perhaps the most interesting is the "Parliament Oak," which was once included within the boundaries of Clipstone Park; it stands by the side of the road leading from Ollerton to Mansfield, and its massive trunk, now rent and shattered by time, is carefully guarded and supported by timbers placed around it through the praiseworthy care of the late Duke of Portland. The trunk is so much decayed that it is cleft into two distinct portions, the top of each being terminated by a goodly arm in full vigour, and in most years bearing acorns in abundance. I used to wonder at the vitality of this old tree, cleft as it is to the ground through the centre, the whole upper part gone, and the heart of what remains decayed away, until a mere shell is left, but on examination I saw the secret of its verdant appearance. The bark which once clothed the shattered trunk, has gradually decayed and fallen off, leaving only a strip extending from the base of the arm to the ground. Up this channel the sap has flowed until the once flattened strip has swelled so much that it looks as though the rounded stem of a tree about the size of a man's thigh had been placed upright against the bare trunk to the base of the arm. It would thus be possible to cut the whole of the trunk away without affecting the vitality of the arm.

Many a traveller who knew not its history would pass this venerable tree unnoticed, but mere relic as it is, it once spread its sturdy branches over the head of royalty, and it happened in this wise.

During John's residence at Clipstone in 1212, he and his barons were one day hunting in the forest, when a messenger in hot haste sought the royal party, bringing intelligence of the second revolt of the Welsh. The king summoned his lords around him to a brief conference under the spreading branches of a huge oak, which from that day has in consequence borne the name of the "Parliament Oak." The result of this hurried council was that orders were at once sent to Nottingham to execute the Welsh hostages there confined in the castle.

This remarkable tree has been computed to be from a thousand to fifteen hundred years old. The following were its dimensions when I last measured it, although it is for age and not for size that it is noted. Circumference on the ground, twenty-seven feet seven inches; ditto fourteen feet from the ground, thirty-two feet six inches; extent of branches from the trunk, sixty feet. The internal diameter varies from three, to eight or nine feet. There are some other trees whose celebrity, though not historical, is still well deserved. The "Greendale Oak" in Welbeck Park is one of these giants, and can boast of an age little inferior to its royal brother. So great was its diameter that a former Duke of Portland cut a *carriage way* through the trunk. Evelyn and Strutt both give figures of the tree, and the latter says in his Sylva, "In 1724 a roadway was cut through its venerable trunk higher than the entrance to

Westminster Abbey, and sufficiently capacious for a carriage and four horses to pass through it." The dimensions of the archway were, height ten feet three inches, width six feet three inches. In consequence of the shores by which the old tree is supported, the feat of driving through it cannot now be performed, but I have often walked through, and wondered that it still retained any life.

Welbeck Park has produced several other oaks remarkable for size. One called the "Duke's Walking Stick," which was cut down about 1800, was one hundred and eleven feet, six inches high, the first branch springing out of the trunk at the height of seventy feet, six inches; the circumference of this tree at fourteen feet from the ground was fourteen feet. There is another worthy to take the place of this fallen giant, called the "Young Walking Stick," of the juvenile age of 140 years; it is more than one hundred feet in height, and so straight and clean that it would be nearly fit for the mast of a ship as it stands, though its circumference at three feet from the ground is only five feet.

Thoresby Park also contains some fine oaks which would almost rival the one just mentioned, whilst there are scores of others which elsewhere would be noticeable, but here in the presence of their loftier companions they attract little attention except from the few genuine admirers of nature. Rufford Park, too, can boast of some exceedingly lofty beeches, though I regretted to notice on a late visit that several of the largest had lost one or two arms from the effect of high winds, which detracted from their general symmetry.

Leaving the parks we will turn into the open forest, and there amongst thousands of venerable old trees, scarcely one of which but is more or less shattered by the storms of nearly a thousand years, are many worth particular notice. The most remarkable of these is the "Major Oak," which stands by the side of the broad riding in Birkland, and, unlike most of its fellows, it is externally unharmed; it is indeed a noble tree, and covers with its spreading boughs an area of 240 feet in diameter. It is not, however, until the spectator approaches closely that the enormous bulk of both trunk and branches is realized, several of the arms are each of them large enough for a tree of no mean size. The massive trunk, however, which looks so vast and firm, with its huge cable-like roots anchored in the ground, is hollow, the cavity being entered by a cleft only sufficiently wide to admit a person sideways. But though the entrance is narrow, the interior is large enough to contain twelve adults standing closely together, and I have seen twenty-seven school-children stowed at once into its spacious recesses; indeed on one occasion I remember a party of four actually sitting down to tea within the tree, with a small round table in their midst.

The "Shambles Oak" in Bilhagh is of almost equal proportions, and owes its name to the fact that a notorious sheep-stealer, who once resided in the village of Clipstone, used its spacious hollow trunk as a place of concealment for the sheep he had stolen and slaughtered. Hundreds of other trees, unknown to fame by any traditional remembrance, are yet worthy of all admiration for their picturesque forms and venerable appearance.

Although Sherwood Forest is royal no longer, no part of it now belonging to the Crown, yet it was for some centuries one of the sources from which was drawn a large portion of the timber used in the construction of our "wooden walls," now, alas! becoming things of the past.

At various times portions of the forest had been granted to different lords of manors, the last which changed owners being that beautiful tract adjoining the little town of Ollerton, comprising the two Hays, or divisions, of "Birkland" and "Bilhagh," still covered with trees. This about sixty years since was granted to the late Duke of Portland in exchange for the perpetual advowson of St. Mary-le-Bone, then held by the duke. The division of Birkland (so called from the numerous birch trees which are interspersed amongst the oaks), was reconveyed by the duke to the late Earl Manvers in exchange for the manors of Holbeck and Bon Busk, which were contiguous to the duke's domain at Welbeck —Thoresby Park, the seat of the earl, being only separated from the forest by a fence.

Some idea of the value of the timber derived from the forest in bygone times may be estimated from the fact that in 104 years—viz., from 1686 to 1790—there had been cut down no fewer than 27,199 oak trees, all of large size. From a survey made in 1609 it was found that even then the majority of the trees were past maturity. The result of the survey was that Birkland numbered 21,009 oak trees, and Bilhagh 28,900. I have heard my father say he remembered that in each of the three consecutive years before the Crown parted with the forest, a large fall of trees was taken for the royal dockyards.

Some curious instances of inosculation occur amongst the trees in Thoresby Park, but this peculiarity does not seem to be possessed by all trees even of the same

UNIV. O
CALIFOR

INOSCULATED SCOTCH FIR.
Thoresby Park.

species. On the eastern boundary of the park there is a long row of fine trees, chiefly beech and lime, and in one of the former this tendency is strikingly shown, for there are at least twenty places where the boughs have crossed each other and become perfectly united. In some of these the opening formed by the junction is only large enough to receive the hand, whilst in others a person might almost creep through; indeed it seems as if two boughs could scarcely touch each other without uniting.

I have seen this tendency exhibited in the beech and elm, though I never met with it in the oak, and but once in the Scotch fir; this last, however, is a very remarkable example, for it is not the mere junction of two boughs of the same tree, but the actual union of two distinct trees. These trees stand in a wood called the "Catwins," not far from the road leading from Thoresby to Clumber. The largest grows with a straight clean trunk about two feet in diameter, while two feet from it grows another smaller tree of equally clean growth of twelve inches diameter. At the height of about four feet the latter bends at an obtuse angle towards the larger tree until at six feet from the ground it touches it, when the two form a perfect junction, combining into one straight, round stem, without any seam to indicate the point of union. The whole has a very singular appearance, the smaller tree looking exactly like a flying buttress placed to support the other.

In some tracts of the forest the oaks are replaced by thorn trees of great size and age, and almost every one bearing in abundance that peculiar parasite the mistletoe; plentiful as it is on the hawthorn, I never met with it on the oak, and perhaps it may have been its rarity on the latter tree that gave it sanctity in the estimation of the Druidical priests.

I cannot enter into the botanical treasury of the district, but it is rich in rare plants, and especially in mosses.

It is hard to say when the forest is most attractive, though each season has its peculiar charms. It is certainly gayest in the spring, when the gorse with its yellow bloom is mingled with the rich green of the fern, and the glowing purple of the heather; while the wild cherry, the mountain ash, and the hawthorn enliven the woods with their fragrant blossoms, and many a "bank whereon the wild thyme grows," is rendered brighter with the graceful harebell, the yellow vetch (*Vicia lathyroides*) with its scarlet buds, the dog violet, and a host of others. With the approach of autumn the summer flowers have departed, but the woods are steeped in a multiplicity of colours, blending into each other with marvellous richness. Nor is even winter without its special beauties. No one could walk in the depth of the forest when, after a misty night accompanied with frost, every twig of sturdy oak and graceful birch, every sprig of heather, or frond of withered fern, or stalk of dry but feathery grass is clothed with particles of ice, which glitter in the rays of the sun like diamonds, without owning their admiration for a scene so fair.

With the localities I have thus slightly sketched I have been familiar from childhood. My home was within ten minutes' walk of the forest, and as I could there roam about with unrestrained freedom, it was my favourite resort. Boyhood increased my attachment to the wild scenes, and kindled a taste for natural history which to the present day has been one of the sources of my purest earthly pleasures. It became my delight to watch the habits of the animals and birds I met with in my rambles, until there were few that were not familiar to me. Like most youths I had a spice of romance about me, and I must confess that in spite of the stern realities of life, it lingers still. Hour after hour have I wandered, sometimes with a companion, but more frequently alone, in all parts of the forest by night as well as by day, until I almost came to know each individual tree, and to look upon them as old friends. With what delight have I watched on a summer's eve the glow-worms light their lamps literally by thousands, until almost every blade of grass and frond of fern bore its tiny beacon fire. Delighted have I sat

"To listen as the night winds crept
From leaf to leaf,"

and as darkness shrouded everything from view I have derived an inexpressible pleasure from the various sounds which fell on the ear.

The long continued *whirr* of the night-jar would alternate with the hoot of the white, or the screech of the tawny owl, or the wild cry of the stone plover would ring out clear as it passed overhead; while in some seasons, but not always, the nightingale would make the woods echo with her song.

Year by year the habit and love of observation has grown on me, and one of the results has been the following notes on our forest birds.

Under this title I have included all I have met with in the district, both constant residents and regular or occasional visitors, the total number amounting to 172 species.

CHAPTER II.

BIRDS OF PREY.

THE woody district described in the preceding chapter possesses two great attractions for rapacious birds—viz., shelter and food, combined with a large amount of seclusion. Minor matters may more or less influence the occurrence of a particular family or species in a locality, but it is the all-important question of food which determines their greater or less frequency; the abundance of game and waterfowl which are strictly preserved on the numerous domains around us, offers to birds of prey an additional attraction to their ordinary supplies, and, as might be expected, a large amount of black mail is levied by these winged reivers for their own especial use.

In one respect, indeed, abundance of game might be considered unfavourable to the increase of birds of prey, for in Great Britain game is invariably accompanied by gamekeepers, and *they* hold the whole tribe in detestation, and lose no opportunity of showing their hostility to such "varmint." This is sadly testified by the numerous victims whose bleaching skeletons adorn the doors and walls of the feeding hovels, and other similar places. Yet, in spite of the war which is so unrelentingly waged against them, the causes I have previously mentioned combine to prevent their extermination, or ere this, in England at least, the whole family would have been extinct.

As forcibly illustrating this point, I copy from an *Inverness Courier*, of May, 1856, the following account; it is headed, "Extraordinary Destruction of Vermin," and extraordinary indeed it is. The list has not been made by a naturalist, and there is consequently some confusion as to the precise species meant; but there is nevertheless the plain fact, that on one estate, in four years and a half, 818 individuals belonging to the order under consideration were destroyed by the keepers. But here is the account :—

"The Marquis of Ailsa has for some years encouraged his gamekeepers in the destruction of vermin by paying so much per head for those brought in. Every keeper and assistant-keeper has a record of all the vermin killed by him, and he receives payment every three months accordingly, besides the regular and liberal wages to which they are entitled for their services. All kinds of vermin were thus brought low, even to the jackdaw and common rat, which, we are informed, caused great destruction to the eggs of pheasants and partridges. The rat has become very common there, and is found to burrow in rabbits' holes to such an extent, that in ferreting rabbits, it sometimes happens that a rat and a rabbit are shot right and left. Whole broods of young pheasants and partridges have been found dead, and partly eaten near rats' holes, and sometimes even young hares and rabbits. The owl, generally supposed to be harmless, has been shot with young game in its talons; and hedgehogs have been found with large accumulations of eggshells in their burrows, or in the long grass where they coil themselves up, and have always been taken in traps baited with rabbits, hares, or woodpigeons. Adders are included amongst the vermin to be destroyed on the marquis's property; but we believe they do not injure game. In four years and a half, the sum paid on the estate of Culzean and Craiglure moors, in Ayrshire, for vermin destroyed by the keepers, amounted to 231l. 15s. 10d."

The following is a list of the vermin of all kinds killed from June 25, 1850, to November 25, 1854:—

Foxes	32	Ash-coloured Hawks	310
Otters	19	Kestrel Hawks	231
Badgers	1	Merlin Hawks	7
Cats	1296	Small-Horned Owls	123
Martin-cats	2	Large Glede Owls	113
Pole-cats	43	Common Fern Owls	33
Stoat Weasels	2132	Yellow Barn Owls	21
Common Weasels	1942	Ravens	52
Rats killed in woods and hedges	12,586	Hoody, or Carrion Crows	225
		Magpies	395
Hedgehogs	1093	Jaypies	4
Adders	267	Jackdaws	1041
Falcon Hawks	6		
Buzzard Hawks	7	Total	21,981

It is certain that the wholesale destruction of animals which act as a check to the undue multiplication of other classes of life, cannot fail to operate in many ways injuriously to the welfare of the landowner and the agriculturist. A proprietor like the Marquis of Ailsa may, by such an indiscriminate slaughter, secure a slight increase in the number of partridges and pheasants on his estates; but nature's equilibrium cannot be disturbed with impunity, and perchance he little dreams how deeply his own pocket and that of his tenants are touched by such a course. The woodpigeon, the field-mouse, the cockchafer, and the wireworm, may be specially instanced as seriously affecting by their ravages both the corn crops and young plantations; but remove the checks which keep them within the bounds designed by their all-wise Creator, and they rapidly increase, the damages they commit being multiplied immensely, and possess a money value which few who have not examined into the matter, would calculate upon. Instead, therefore, of a supposed gain being derived from a practice, which I fear is too prevalent, an actual loss is the general, and I believe inevitable, result.

This subject was incidentally alluded to by Sir William Jardine at the meeting of the British Association in 1856, in connexion with the artificial propagation of salmon in the Tay, a subject on which he was specially deputed by his section to report. In the course of his remarks, he stated that it had been found that one of the worst enemies of the salmon ova in the breeding beds was the larva of the Mayfly, which, in its turn, was a favourite food of the trout. Now the practice in rivers preserved for salmon-fishing was to destroy trout, while this fact clearly showed that such ought not to be the case, as, by keeping down the Mayfly, they aided in propagating the salmon. As an illustration of this immutable law of nature, Sir William pointed out that in parts of the country where hawks had been ruthlessly extirpated, with the object of encouraging the breed of game, woodpigeons had increased to such an extent as to have become a positive nuisance, and most injurious to the farmer; and he showed the danger incurred by unduly interfering with the balance established by nature amongst wild animals. But I shall recur to this point again.

The rapacious birds usually considered as British amount to thirty-three, and of these I can number nineteen as having come under my notice as rangers or visitors in "merrie Sherwood."

The first is the Golden eagle (*Aquila chrysaëtos*). I was aware of a reported visit of this noble bird, but as I had not seen it myself, and could not authenticate its occurrence, I determined to omit all mention of it. A recent letter, however, from Mr. Tillery of Welbeck, to *The Field* of January 27, confirms the report I had heard, and enables me to include it in my list.

It was in the winter of 1838 that the bird appeared in Welbeck Park. Mr. Tillery says:—

"The lake was frozen over at the time, except in one place, where a flush of warm water entered from a culvert which drained the abbey. The place was covered with ducks, teal, and widgeon, and I saw his majesty swoop down once or twice to get one for his breakfast, but unsuccessfully, as the ducks saved themselves by diving or flying off. The park-keeper got two shots at him with ball on a tree, but missed him each time, and he gradually got wilder, so that he could never be approached again near enough for a shot. After levying black mail on the young lambs, hares, and game in the neighbourhood, he took himself off after a three weeks' sojourn."

I am enabled, through the kind attention of a friend, to add two individuals of another species, and that one of rare general occurrence in England—viz., the White-tailed sea-eagle (*Haliæetus albicilla*) to my list. This eagle appears somewhat subject to a partial southern migration in the winter, and it has been usually at that season that it has been noticed in England. It is also rather discursive, frequently proceeding some distance inland from its usual haunts on the sea-shore. The birds in question were no exception to these peculiarities, one being shot at Osberton, a few miles from Ollerton, at the beginning of January, 1857, and the other was killed several days later (January 13, 1857), at Lina Wood, near Laughton-en-le-Morthen, just across the northern border of Nottinghamshire.

The latter bird was seen in the neighbourhood of Morthen for more than a fortnight before it was shot. On several occasions it was observed perched in a tree about a hundred yards from Pinch Mill, the person resident there taking it at that distance for a stray heron. Thomas Whitfield, the gamekeeper to J. C. Athorpe, Esq., of Dinnington, made many attempts to get within range of the bird, but was as often baffled by its wariness. It was observed to be much molested by crows and small birds, and frequently, as if to escape from persecutions which were beneath its notice to resent, it would mount into the air with graceful spiral curves until it became nearly lost to sight, leaving its puny assailants far below, and then would sweep as gracefully down again, with all the ease and lightness of wing of the swallow.

It seems uncertain what its food consisted of during its sojourn, for it was not seen to make any attack. At night it roosted on a tree, but still maintained a vigilant watch. When perceived by Whitfield, it was perched on a tree on the outskirts of the wood; but the night being moonlight, it perceived his approach, and he had great difficulty in getting within gunshot. At the moment of his firing it flew off, and he thought he had failed in hitting it; but in the morning he found it dead in an adjoining field. Its expanse of wing from tip to tip was seven feet six inches, and its weight eight pounds and a quarter.

The friend I have mentioned kindly procured the loan of the bird from Whitfield, and sent it for my inspection. It is a fine specimen, in the immature plumage of the third or fourth year, and its markings are as follows:— Hackled feathers of the head and upper neck dark brown, the basal portions being white, which here and there shows through; throat, breast, and back different shades of white and brown, somewhat mottled, each feather having an oval brown tip; back, belly, shoulders, and upper wing coverts mottled, brown and white; thighs brown; primaries and secondaries dark brown, the tertiaries being rather lighter; upper tail coverts mottled, brown and white, lower tail coverts dirty white, the terminal portion of each feather being edged with dark brown, and distinctly tipped; tail feathers, inner webs dirty white, outer webs brown, somewhat mottled on the centre feathers; bill dark horn colour; legs yellow, with formidable black claws.

The next is one which I am afraid is rapidly becoming scarce even in its own Scottish fastnesses, the Osprey (*Pandion haliæëtus*), one having taken up a temporary abode on the borders of the lake in Thoresby Park in the summer of 1855. I have only once noted the previous occurrence here of this fine fish hawk; indeed, it is very rarely seen so far from its usual haunts, but the one in question must have been attracted in its wanderings by the piscine resources of the large sheet of water where I had the pleasure of seeing it. Here it remained some weeks, faring sumptuously, its manner of fishing affording me and others who witnessed it much gratification; its large size, its graceful manner of hovering over the water when on the look-out for its prey, and the astonishing rapidity of its plunge when darting on its victim, rendering it a conspicuous object. It then to my great regret took its departure, doubtless alarmed at the attacks of the keepers, who viewed its successful forays with little favour. The other was shot by a keeper on Welbeck Lake a few years before.

Foremost among the typical falcons I am glad to include the noble Peregrine (*Falco peregrinus*), the very perfection of a bird.

By the uninitiated observer in most rural districts, all large birds of this kind are classed under the cognomen of "hawks," without discriminating one species from another. The falcons, the buzzards, and the harriers are all undistinguished from each other by country people when seen in the air; and I am induced to believe that the peregrine is a far more frequent denizen of our forest than is generally supposed, for I have noted the occurrence of four individuals in the course of five years. Of course they do not breed with us, for we have none of the cliffs and headlands on which it delights to place its eyrie, but a flight of a few hundred miles is nothing to a bird whose speed rivals or excels the best efforts of the locomotive, and a journey to or from its distant home is soon performed. The keepers say they know it, and call it the "blue hawk," although the male of the hen harrier is generally known by this name; but it is so much more frequent than the latter species, and differs so greatly in its flight and general appearance on the wing, that few who really knew the two could mistake them; yet, strange to say, with all their opportunities, keepers are not to be depended upon for the identification of a species.

I know few more beautiful sights than to watch the flight of this noble bird, especially when it is in the pursuit of its prey. Slowly winging its way on high, as though in the mere enjoyment of its leisure, its keen eye marks a mallard rise from the rushy stream, or a woodpigeon from the stubble, and instantly the graceful curves of its flight are arrested, and with quick strokes of the wing it rushes with the speed of an arrow in pursuit of its destined victim. Woe betide the unfortunate quarry, for its chances of escape are small; even the rapid flight of the woodpigeon avails nothing against the headlong rush of the peregrine. In a few moments the distance between the pursuer and the pursued is fatally diminished, when, with a velocity of which it is almost difficult to conceive of the flight of a bird unless witnessed, it swoops upon its victim, dashing it to the ground with a stroke of its formidable hind claw, and immediately rising obliquely in the air it checks the impetus of its course, and then returns to pick up its prey.

What is the maximum speed of the peregrine's flight? On this point I was appealed to by the gentleman who, under the pseudonym of "Peregrine," has written much on the ancient sport of falconry in the columns of *The Field*, and who is one of our best authorities on the subject. After admitting that the best estimate must of necessity be but an approximation, I expressed my belief that the actual speed attained by this noble bird during the death rush was at the rate of 150 miles an hour! Many who have not witnessed it, may be incredulous as to any bird flying with such velocity, but "Peregrine's" long experience as a falconer with birds of this very species enabled him fully to confirm my estimate, though, as he says, it is difficult to conceive the lungs of any bird lasting through it. Of course this rapidity of motion could not be maintained for long, but that it is actually attained I have not the shadow of a doubt.

It was a strange idea that the peregrine killed or disabled its prey by a blow with its breast; but this is such an evident absurdity that it has long been discarded. Alexander Wilson thus alludes to this opinion:—

"From the best sources of information we learn that this species is uncommonly bold and powerful, that it darts on its prey with astonishing velocity, and that it strikes with its formidable feet, permitting the duck to fall previously to securing it. The circumstance of the hawk never carrying off the duck on striking it, has given rise to the belief of that service being performed by means of the breast, which vulgar opinion has armed with a projecting bone adapted to the purpose. But this cannot be the fact, as the breast-bone of the bird does not differ from that of others of the same tribe, and would not admit of so violent a concussion."

In Montagu's Ornithological Dictionary,* an account is quoted from a writer in a popular periodical of a peregrine pursuing a razor-bill, and stating that "instead of assaulting, as *usual*, with the death pounce from the *beak*, he seized it by the head with both his claws." This "as usual," is undoubtedly a mistake, for the stroke is given with the *foot*.

Byron was no better ornithologist than the writer

---

* Second edition, 1831.

above quoted, for in the original draught of "Childe Harold" (canto 3, stanza 18), he wrote:—

"Here his last flight the haughty eagle flew,
Then tore with *bloody beak* the fatal plain."

In the edition of 1837 this note is appended:—

"On seeing these lines, Mr. Reinagle sketched a spirited chained eagle, grasping the earth with its *talons*. This circumstance being mentioned to Lord Byron, he wrote thus to a friend at Brussels: 'Reinagle is a better poet and a better ornithologist than I am; eagles and all birds of prey attack with their talons, and not with their beaks, and I have altered the line thus:—

"Then tore with bloody talon the rent plain.

"This is, I think, a better line, besides its poetical justice."

On April 5, 1856, I obtained a peregrine, which had just felled a common blue pigeon to the ground by a stroke of its foot, and was descending to pick up its prey, when it was shot. It was a remarkably fine male bird in perfect feather, with the markings clear and distinct. The pigeon, when taken up, was found to have received a fearful blow, the back being deeply ripped up, clearly evidencing the force with which it had been struck.

Of the smaller falcons, the hobby and the merlin occur with us at short intervals, whilst the kestrel is as abundant here as it seems to be in every other part of the kingdom.

Although the Hobby (*F. subbuteo*) is always considered a summer visitor, yet it is strange that I have never met with it at that season. My note-book records the occurrence of three individuals in about eight or nine years, but all of them were in winter—viz., one in November, one in December, and another in March; the latter might be accounted for as a particularly early arrival, as it was a fine adult male; the other two were in immature plumage, and may have been hatched late, but I believe it is quite exceptional to meet with them in mid-winter.

I have no knowledge of its breeding with us, and have in vain sought for its nest; and yet the fact of meeting with birds in the livery of the young is pretty strong proof that it does so, though it has escaped my notice. At the same time, no country could be more favourable to its habits, or furnish a more abundant supply of food. It is not unlikely that the incessant attacks of the keepers upon everything in the shape of a hawk may have taught it instinctively to seek safer quarters.

The little I have seen of the hobby has impressed me with the opinion that, though naturally wild and shy, yet when in pursuit of its prey it will be absolutely fearless even of man's presence, and will pertinaciously follow on regardless of any danger to itself.

The beautiful little Merlin (*F. æsalon*) is more frequent than the hobby, and though it cannot be called rare, it is sufficiently so to make it an object of interest when seen on the wing. Like the latter, too, most of the occasions on which I have met with it have been during the winter months, several having occurred in February. It haunts a wild heathy tract at Inkersal more commonly than elsewhere; and here, with little to disturb it, it breeds, placing its nest on the ground, which is, I believe, its usual practice. I have often thought the latter circumstance must, even in districts which merlins habitually inhabit, account in some degree for their scarcity in comparison with the kestrel, their eggs and young being destroyed by weasels, rats, and other predatory animals; were it not so, I think we should find them more abundant, for both lay the same number of eggs, their food is equally plentiful, and when they are able to fly they are equally liable to the attacks of the same enemies; yet, after all, the kestrel is a hundred times more abundant than the merlin.*

The merlin is a compact little bird, and, in fact, might be called a miniature peregrine, for it fully equals it in boldness and spirit; though it does not secure its prey with the same dashing flight, yet it is quite as unrelenting in pursuit, and even more persevering, following it in all the efforts it makes to escape. It is not at all particular as to the size of its victim, a partridge or a linnet being pursued with equal ferocity, and I have known one shot whilst killing a skylark, its most favourite food.

Every part of the district is frequented by the Kestrel, (*F. tinnunculus*), especially the more heathy parts of the forest, and it would be difficult at any time to go far without seeing one poised buoyantly on outstretched wings, with head depressed, and eyes eagerly scanning the ground below. It is to be deeply regretted that a bird which is at once an ornament to the landscape and a benefactor to the agriculturist should be so constantly persecuted. I have often tried to reason a keeper into the fact of the kestrel and the owl being his best allies

---

* This discrepancy in the respective numbers of allied species is one of those facts we cannot account for, and can only acknowledge that it is so.

in the destruction of various kinds of vermin more or less injurious to game; but it has been a fruitless task, and one instance which he may be able to adduce of either one or the other having dined or supped on game (for I do not deny that they sometimes vary their diet in this way), fairly outweighs with him all their good qualities, and makes them outlaws at once.

Their labours are, indeed, highly beneficial to the farmer and the forester, their chief food being field mice, of which they destroy vast numbers, and various kinds of beetles, especially the cockchafer; birds are sometimes preyed upon, though I believe not so frequently as by others of the family—at least, I have observed that the kestrel is less subjected to the noisy pursuit of small birds, and I have often watched it hunting where no notice of its presence has been taken.

It is plentiful with us throughout the year, although I have no doubt that it is partially migratory, as we certainly receive an accession of numbers in the spring, which again gradually leave us in the autumn.

I have known the nest placed in the hollows of some of the blasted tops of the old oaks, and sometimes a deserted nest of the crow or jay has been made use of. I have an egg in my collection which, instead of being blotched and marbled in the usual manner, is of a uniform chestnut colour.

What I have said of the kestrel cannot, however, be applied to the Sparrowhawk (*Accipiter fringillarius*). Its appearance is the signal for a general commotion, and the cries of alarm which are uttered when first the presence of the enemy is detected are well understood by various species, and swallows, linnets, chaffinches, and others, instantly muster at the summons and join in hot pursuit.

I have noticed on these occasions that linnets and chaffinches usually follow in steady chase, while now and then one more bold than his fellows will dart forward and make a momentary attack on the hawk, and then rejoin his companions; but the swallows, with much greater power of wing, fly wildly to and fro, now darting across his path, then shooting ahead, and again returning, all the time uttering cries of fear and hostility. The sparrowhawk seems generally to hold all his noisy persecutors in supreme contempt, excepting that now and then his patience becomes exhausted, and with a fierce sally he sacrifices one of them to his resentment. In this it shows more spirit than the kestrel, which I have often observed to be apparently annoyed, and more anxious to escape such boisterous recognition than to become the aggressor.

I have often remarked how quickly the poultry in my yard have caught sight of a kestrel or a sparrowhawk on the wing. The watchful cock is generally the first to utter his warning cry, which is immediately repeated by the hens, and all, with head turned sideways, scan the course of the intruder, those who have chickens instantly calling them ogether for protection until the danger is past.

We hear from all parts of the country of the large amount of damage done to the crops by woodpigeons. In some districts of England and Scotland meetings have been held to devise means for their destruction, for they have enormously increased of late years. We cannot wonder at this, for their natural enemies are unrelentingly extirpated, and the sparrowhawk is especially a foe to the woodpigeon, preferring it to any other quarry. During the first week in the present year (1869) a flock of these birds passed over Barnet, which was estimated to be a mile long, and to contain from 8000 to 10,000 in number.

The sparrowhawk is not such an abundant species as the kestrel, though still very common. It has always struck me that the females are much more numerous than the males, though I am uncertain if it is not more apparent than real, the greater boldness and spirit of the former bringing them more frequently into notice.

The Goshawk (*F. palumbarius*), rare in Scotland, though said to be resident there, is still rarer in England. I never saw the bird on the wing, and only once in the flesh, and we seem to know very little of its life history. Rare as it is, a single specimen was killed by one of the keepers near Rufford in 1848, being the only instance I have known of its occurrence, and I am thus able to add it to my list.

The Kite (*Milvus regalis*) is now with us, as it is elsewhere in England, a comparatively rare bird; I have noted several instances of its occurrence of late years, two coming under my own immediate observation, and one of these under circumstances of much interest. I was riding on horseback over the forest between Ollerton and Budby, where it is open and heathy. On a strip of greensward, bounding the unenclosed road by which I was proceeding, a brace of partridges were busy searching some droppings of dung. I had approached them within about some fifteen yards, when a kite glided across in front of me, and made a swoop at the partridges on the ground; but, whether his aim had been rendered unsteady by my close approach or not, he missed them both. The poor birds with loud cries of alarm scudded along for a short distance, and then dropped into the long heath; while the kite, urged onwards by the impetus of his swoop, rose gracefully in the air, and wheeling over my head, mounted upwards in spiral circles, gradually increasing in circumference until he became a mere speck in the blue sky, and was lost to my gaze. I could not help admiring the ease with which this was done, for after he commenced his gyrations I could not detect a single stroke of his wings; he soared aloft as if propelled by some invisible power, his forked tail, broadly outspread, alone moved, as rudder-like, it directed his graceful course.

Another, a female, was shot near Edwinstowe while flying from the old trees of Birkland to the grounds surrounding the house of the late Dowager Countess of Scarborough. At one time the kite was by no means uncommon, but, from its habit of taking its prey on the ground, it is easily trapped, and I have little doubt that this is one of the chief causes of its diminished numbers.

To me there is something very wild in the shrill squeal of the kite, and when heard in the moorland districts, where now it is chiefly to be encountered, it has a peculiarly drear and mournful sound.

I have met with all the buzzards more or less frequently, especially in the wilder and less wooded parts of the district; but I have not found any of them breeding with us. Those whose presence I have recorded must have migrated from other quarters, and they, alas! are no sooner seen than they are picked off by the keepers, leaving their places to be supplied by fresh victims.

The Common Buzzard (*Buteo vulgaris*) is the most frequent species, and indeed may be called common. It is more discursive in its habits than the other two, but appears to prefer the neighbourhood of the older woods and plantations, while the bare heathy tracts which approach the moorland in character are the choice of the rough-legged and honey buzzards.

The common buzzard appears a much larger bird when on the wing than it really is, especially when seen soaring, as it sometimes does, at a great elevation; but when hawking for food it flies at a very short distance from the ground, and with a peculiar gliding, noiseless flight. It seems especially fond of beating along the outside of plantations which are surrounded by hedges, pouncing on the rabbits which sally out in the evenings to feed, and which form its favourite repast.

Inkersal Forest, about three miles south-west of Ollerton, is a favourite haunt of the Rough-legged Buzzard (*B. lagopus*), several having been there met with during the last few years. I obtained a very fine male which was shot there by a keeper in 1857 while on the wing. The shot took effect and brought him down, but, as it afterwards proved, he was not seriously hurt, excepting that one of his wings was broken. As soon as the bird dropped on the heather a retriever, accompanying the man, started off at once to fetch him in, but no sooner had the dog approached, and was about seizing the buzzard in his mouth, when, doubtless much to his astonishment, the bird of which he expected to make so easy a capture at once sprang on his back, and began to ply his beak and talons in good earnest. For a short time the unequal contest continued, the poor dog in vain attempting to free himself from his antagonist, who inflicted on him some severe wounds before he could be dislodged by the keeper from his post of vantage. The plumage of this bird was altogether of a much darker tone than is usually the case, the brown markings on the centre of the breast and belly being very deep and large, as well as those on the feathers of the thighs and legs.

I have met with three instances of the occurrence of the Honey Buzzard (*Pernis apivorus*). One was killed at Rufford in the summer of 1854; and a pair, male and female, were taken together in the same trap in Inkersal Forest, on April 26, 1858. This was a remarkable capture, for the birds, with singular unanimity, whether caused by hunger or rivalry I cannot tell—most likely the former—must have pounced on the rabbit with which the trap was baited at the same moment, for both were caught by the legs, and of course forfeited their lives. Their seizing on the rabbit is a proof, too, that this species does not confine itself to insect diet; and in the absence of the larvæ of bees and wasps I imagine it is not very fastidious.

The circumstance of these birds having mated, and the time of year when they were taken, are strongly presumptive that they would have bred here had they not met with such an untimely fate, for the locality is a very lonely one, and offers many suitable spots for a nest.

The harriers are not nearly so plentiful with us as in many other parts of England, and especially of Scotland. Our district is not generally favourable to the natural tastes and habits of the Marsh Harrier (*Circus rufus*), and I have only noted the occurrence of one; but the Hen Harrier (*C. cyaneus*) is frequently met with, although individuals in brown plumage are much more numerous than the grey-plumaged male. This is, I think, easily accounted for. These brown birds, which are known as ringtails, are generally considered to be females, yet such is not the case, the young males wearing the livery of the female until they are a year old, and not being easily distinguished without dissection. Still, the "blue hawk," as the male is called, is not by any means uncommon; and both male and female being considered, and I fear not unjustly, as very destructive to game, are visited, whenever opportunity offers, with condign punishment, and their once buoyant forms are often seen nailed up *in terrorem* amongst others of their order, in grim companionship with stoats, weasels, polecats, and other vermin.

The ringtail, as is usual in birds of prey, is much larger than her mate, and far bolder in her search for prey, not hesitating to frequent the neighbourhood of dwellings for the chance of picking up a stray chicken. In 1857 I was walking past Lord Manvers' poultry-yard at Perlthorpe, which adjoins Thoresby Park, when a ringtail came sailing over, evidently intent on plunder. Three times she soared round the large inclosure, which contained several hundred head of poultry, and although it is bounded by a high wall, and surrounded by the dwellings of the keepers and others, she was only deterred from carrying off a fowl by the presence of some of the men.

All parts of the forest are frequented by them, though they seem to prefer the more open portions. Here, however, their habit of seizing their food on the ground often leads to their destruction, as they are easily trapped, and I have had them frequently brought to me when thus taken.

I have been surprised to find some writers expressing a doubt as to the propriety of applying the name of harrier to birds of this genus. I cannot but think they well deserve the appellation; none others of the family obtain their food in the same manner, and few who have carefully witnessed their mode of hunting will doubt the correctness of Mr. Yarrell's suggestion that "the origin of the name has probably been derived from their beating the ground somewhat in the manner of a dog hunting for game." They do not spy out their prey from a distance, or pursue and strike it in the air, as is the habit of the falcons, but with a low, buoyant, but somewhat stealthy flight, they steadily hunt and quarter the ground, as a well-trained pointer might do, until they catch sight of some unfortunate bird or rabbit, when, if practicable, they pounce upon it at once, or, if it take the alarm, they chase it until it is secured. The meaning attached to the word "harry" in the dictionaries—viz., "to tease, to hare, to make harassing incursions," is so descriptive of their habits that there can be little doubt of the derivation of the name.

Of the nocturnal birds of prey I am fain to be content with a record of four species—viz., the white, the tawny, and the long and short-eared owls. These are tolerably abundant, especially the two former, but I have never seen any of the smaller species, nor am I aware of their occurrence in the district.

An owl is not by any means a popular bird. His grotesque appearance, his wild and unearthly cry, ringing through the air when honest people should be in bed, and his silent, spirit-like flight in the darkness, all combine to invest him with a certain amount of mystery, which in the popular mind does not tend to make him a favourite. His hoot, too, near a dwelling is in some places considered a sign of impending trouble, and, whether he makes his appearance by night or by day, he is always persecuted; yet Minerva's bird ought to meet with better treatment, for, apart from being the symbol of wisdom, the habits and manners of the owl are exceedingly interesting.

THE OWL.

I.

In the hollow tree, in the grey old tower,
    The spectral owl doth dwell,
Dull, hated, despised in the sunshine hour,
    But at dusk he's abroad and well:
Not a bird of the forest e'er mates with him,
    All mock him outright by day;
But at night, when the woods grow still and dim,
    The boldest will shrink away;
        Oh, when the night falls, and roost the fowl,
        Then, then is the reign of the horned owl!

II.

And the owl hath a bride who is fond and bold,
    And loveth the wood's deep gloom,
And with eyes like the shine of the moonbeam cold
    She waiteth her ghastly groom!
Not a feather she moves, nor a carol she sings,
    As she waits in her tree so still;
And when her heart heareth his flapping wings,
    She hoots out her welcome shrill!
        Oh, when the moon shines, and the dogs do howl,
        Then, then is the time of the horned owl!

III.

Mourn not for the owl, nor his gloomy plight!
    The owl hath a share of his good;
If a prisoner he be in the broad daylight,
    He is lord in the dark green wood!

> Nor lonely the bird, nor his ghastly mate:
> They are each unto each a pride—
> Thrice fonder, perhaps, since a strange dark fate
> Hath rent them from all beside!
> So when the night falls, and the dogs do howl,
> Sing ho, for the reign of the horned owl!
> We know not alway who are kings by day,
> But the king of the night is the bold brown owl.

So sings Barry Cornwall, and his spirited lines are very characteristic.

If the constant destruction of the hawk tribe is a matter of regret to the true naturalist, it is doubly to be lamented that the owls are visited with such indiscriminating and ignorant hostility. I will venture to affirm that the good they effect is tenfold, ay, fifty-fold, greater than the injury inflicted by the occasional poaching of a young rabbit or partridge, and earnestly would I raise my voice in their defence, and urge on their destroyers that, even from the lowest and most unworthy motive—that of self-interest—their preservation is desirable.

Bishop Stanley says with great truth: "Generally speaking, a more useful race of birds does not exist, since, with the exception of one or two of the larger and rarer species, their food consists entirely of vermin and insects very prejudicial to our crops, and which, but for these nocturnal hunters, might do serious mischief. A striking instance of their utility occurred some years ago in the neighbourhood of Bridgewater, in Somersetshire, where, during the summer, such incredible numbers of mice overran the country as to destroy a large portion of vegetation, and their ravages might have extended to an alarming degree had it not been for a sudden assemblage of owls, which resorted from all parts to prey upon them." And again : " Some idea may be formed of the number of mice destroyed by a pair of barn owls when it is known that, in the short space of twenty minutes, two old birds carried food to their young twelve times, thus destroying at least forty mice every hour during the time they continued hunting; and as young owls remain long in the nest, many hundreds of mice must be destroyed in the course of rearing them.'

Montagu says, in writing of the tawny owl : " This bird breeds in the hollows of trees, and sometimes in barns, which last it frequents for the sake of mice, and, as it is a better mouser than the cat, the farmer holds it in great estimation, and leaves a hole in his barns and granary for its egress."

I am afraid that on this latter point the farmers now are not so enlightened as in Montagu's day, or does the fault rest with the builders; I have often seen these holes in the gables of old barns, but modern erections are without them. Now the appearance of any species of owl in a farmyard is merely the signal for the production of the gun, and the instant execution of the visitor. I fear this stupid prejudice or practice will retain its sway until a desire to know something of the habits of the various forms which we daily see around us is more extensively diffused than at present, and until the wanton love of destruction is exchanged for a spirit of admiration and reverence for those works which by their divine Creator were pronounced to be "very good."

Perhaps some reader may be inclined to think that, in advocating the preservation of rapacious birds, I have exaggerated the amount of mischief caused by mice, cockchafers, &c., and that the money value of their depredations is not of that extent which I have assumed as probable. Let such carefully read the following authentic account of the destruction of young trees in the Forest of Dean by the short-tailed field mouse (*Mus arvalis*), which was communicated to Paxton's Horticultural Register, by Mr. E. Murphy, and I think they will no longer doubt the value and importance of the checks placed on the inordinate multiplication of creatures apparently insignificant, but which in their aggregate attacks are really so formidable.

After mentioning the appearance and gradual increase of the mice, Mr. Murphy goes on to say : " Before the autumn of 1813 the mice had become so numerous that we could pick up four or five plants of the larger five-year-old oaks on a very small piece of ground, all bitten off just below the ground, between the roots and the stem ; and not only oak and ash, but elm, sycamore, and Spanish chestnut, of which, however, they did not appear to be so fond as of the two former. The hollies which had been cut down produced abundance of suckers, which were destroyed in the same manner; and some of them, which were as thick as a man's leg, were barked all round four or five feet up the stem. The crab-tree, willow, furze, birch, spruce, in a word, every kind of tree, and even grass, particularly cock's-foot grass, seemed equally acceptable to those voracious little creatures, till at length Lord Glenbervie became so alarmed about the final success of raising a forest, that we were instructed to pursue every means we could think of—by cats, dogs, owls, poison, traps, &c. ; but all was to no purpose.

" At length a person hit upon a simple, and eventually a very efficacious mode. Having, in digging a hole in the ground, observed that some mice which happened to fall in could not get out again, the idea of forming similar holes was suggested; it was tried accordingly, and found to answer. In short, holes about two feet long and ten inches broad at the top, and somewhat larger every way at the bottom, were made at twenty yards apart, over about 3200 acres of plantation. Persons went round early in the morning to destroy such mice as might be found in the holes. In this way, besides what the owls, hawks, magpies, and weasels took out of the holes (and several of those depredators lost their lives in attempting to seize their prey), 30,000 mice were paid for by government; nor were they extirpated till they had destroyed 1700 acres, the astonishing number of 200,000 five-year-old oaks, together with an immense number of acorns and young seedlings."

I have frequently met with the Long-eared Owl (*Strix otus*) in the fir plantations in various parts of the forest, these being its favourite places of concealment in the day-time; but it is not common with us, and indeed does not appear to be an abundant species anywhere. It is an amusing bird, and when met with during daylight, perched on a shady bough, it has a most grotesque and perky look; it sits quietly enough if undisturbed, but when roused its ears are instantly erected, and if you put your hand towards it you will quickly experience the sharpness of its beak and claws, with which it fights vigorously.

An allied species (*Strix bubo*), the Eagle Owl, appears to attain a great age, and the keep of Arundel Castle in Sussex is tenanted by some which were introduced many years since by the then Earl of Arundel. One of these died in August, 1859, about the age of 100 years. And it is worth mentioning, as an instance of the length of life attained under favourable circumstances, a fact always difficult to ascertain of wild animals. The bird in question was so noted as to deserve a special obituary by the editor of the *West Sussex Gazette*. He says: "These owls have become almost as famous as the Saxon keep which they occupy. They are very peculiar, and perch up in the niches of the citadel, looking on visitors with a pride which seems to bespeak the dignity of a connexion with the ancient house of Howard. Since their introduction only about six have been added to the family, so that the race is not likely to become common. They usually live to a green old age, but none have ever before passed over a hundred summers. This bird must have been hatched in the reign of George II. Four kings have passed away since it first saw the light, and many Dukes of Norfolk have been numbered with the dead.

"The recently-departed owl was the famous 'Lord Thurlow' of the keep, in connexion with which a ludicrous anecdote is told. It was formerly the custom of the castellan to give each of these birds a name, and from their singularly wise appearance they were invariably named after some celebrated dignitary of the law. One was called Lord Eldon, and the subject of this notice was dubbed Lord Thurlow, we presume in total ignorance of the sex of the bird, which was in reality of the feminine gender. It happened one time that the famous Chancellor, Lord Thurlow, was ill, and much political anxiety was felt at the circumstance. The Duke of Norfolk was desirous of learning the latest intelligence of the learned man, and as he was riding one day into the gateway of the castle, an attendant ran up to him, out of breath, exclaiming, 'Please your grace, Lord Thurlow—' 'Well,' said the duke, sharply, 'what news—is he better or worse?' 'Oh! please your grace,' replied the man, 'he's just laid an egg!' As may be concluded, it is quite an event for an egg to be laid by these aristocratic birds; they do not average amongst them one a year, and it is seldom they are productive."

The Short-eared Owl (*S. brachyotos*) is less frequent than the preceding species, and all the specimens I have seen were in turnip-fields. At the same time, some are met with every winter, and generally in October and November. A male bird was shot on November 9, 1858, while hunting over a turnip-field at Car Brecks, close to the town; it was in most perfect feather, the ears were well developed, and the whole plumage very soft, and, like that of the rest of the family, admirably adapted for a noiseless flight.

This owl is less nocturnal in its habits than its fellows, feeding chiefly, if not entirely, by day; indeed, I never met with it on the move at night, and no amount of sunshine seems to dazzle or confuse it. It possesses great power of wing, though I have always seen it flying near the ground, on which it often alights, and also roosts at night. I think it rarely perches on trees.

Far more numerous with us than any of its congeners is the White or Barn Owl (*S. flammea*), although we have neither ancient towers nor ivy-clad ruins to afford it a shelter or retreat. It is at no loss however, for, in default of these, the old hollow oaks in the forest are generally selected as its breeding places, for which they are admirably fitted, being both comfortable and secure.

In these situations the nest is a mere depression in the decayed and crumbling wood at the bottom of the cavity, without any lining; and so difficult are they of access to any but the owners, that they can only be reached by enlarging the aperture with an axe. Only once have I known of a different situation being selected, and that was an old barn, in which the nest was made of straw, that material being ready to hand.

I believe the white owl is strictly nocturnal in its habits; although I have frequently seen individuals on the wing in the daytime, yet it has been clearly evident that their flight was not a voluntary one, but that their siesta had been accidentally disturbed, for on these occasions they flew in a confused and uncertain manner, as though "blinded by excess of light," and were glad to take refuge in the first tree they met with, manifesting no inclination to leave it unless compelled. These diurnal flights, too, can rarely be made unnoticed, for, like the hawks, they are attended by a numerous following of small birds, who show their hostility by noisy cries of alarm and anger.

I have spoken of the grotesque appearance of the long-eared owl, but that of the white owl is, I think, still more so; when in captivity it wears such an air of mock gravity and wisdom—as though it was intending to burlesque those attributes, now holding its head on one side, and now on the other—that I never look at one without feeling inclined to burst into a fit of laughter, and could fancy it had some difficulty to refrain from doing the same.

I once saw a white owl, which a person had shot and only winged, throw itself on its back when he approached, and fight most vigorously with its sharp claws, rendering it a rather difficult matter to effect its capture without receiving a wound, and all the time fixing its large eyes upon him with a strange weird-like intelligence, but with no appearance of ferocity. I could not help being struck by its expression, as though it was animated by a feeling more akin to reason than instinct, and my heart ached for the poor bird thus struck down so uselessly.

With regard to the note of the white owl, I venture with diffidence to express an opinion which I am aware is contrary to that of most who have written of it; at the same time it is no theoretical fancy, but the result of close and continuous observation. It is generally stated that this species seldom hoots. Montagu boldly says, "it is *never* known to hoot;" Macgillivray, that "it has no other note than a shriek;" and Mr. Waterton, that "the tawny owl is the only owl which hoots." Similar assertions might so easily be multiplied that it seems almost like temerity to assert the contrary. A desire to elucidate the truth, however, compels me to do this, and I am glad to be supported by so high an authority as Sir W. Jardine, who, in a note to a late edition of White's Selborne, says that the white owl does hoot, for he has shot it in the act—and more, that at night, when not alarmed, hooting is its general cry. This I can confirm most unhesitatingly, for I have heard it repeatedly and continuously do so, and, on the very account of the alleged infrequency, have taken particular pains to verify the fact. I have been familiar with both species from boyhood, and have roamed through our forest at all times of the evening, and it was always the tawny owl which we designated as the "screech owl," while the well-known "hoo-hoo-hoo-hooo" was almost invariably uttered by the white owl. I have watched both in various situations, and have often been startled in the woods by the unexpected shriek of the brown owl, while, by blowing into my closed hands, I have imitated a hoot with such exactness as to cause the white owl to approach me very closely. A pair of the latter frequented a small field opposite to my house in the village, and on moonlight nights I have repeatedly and distinctly watched them while uttering their hoot even within a few yards of the house. The result of my own careful and repeated observations may be thus summed up—that the white owl hoots chiefly, but sometimes, though very seldom, screams; while the tawny owl screeches, and rarely, if ever, hoots.

The Tawny Owl (*Ulula stridula*), although not so common with us as the white, is still a plentiful species, our extensive woods favouring its arboreal habits. Its favourite hiding-places in the daytime are the thickly-clothed branches of the spruce and Scotch fir, especially the former. It is not nearly so wary a bird as the barn owl, and if disturbed from its roost before the evening it is even more confused and blinded by the light; but in the dusky twilight it is all activity, hunting on the outskirts of the woods and plantations with noiseless flight, its tawny colour rendering it invisible in the dark shadows of the trees, and I have often been aware of its presence only by its unearthly cry as it glided past me.

The venerable oaks, whose lichened trunks and limbs are rifted and decayed into innumerable cavities, are the places it generally selects in which to breed. In these its young are secure from all enemies except the polecat, and even he would hesitate ere he faced the sharp beak and talons of the old ones, and must be content to make his raids, if at all, in their brief absence.

After what I have previously said, I need hardly add that it is equally an object of persecution with its congeners, and in fact meets with no quarter from its biped enemies.

CHAPTER III.

PERCHING BIRDS.

IN our thickly-wooded and well-cultivated district the birds of this order are pretty numerous, and at least two-thirds of the British species have come under my notice. Of some of these, which are plentiful enough in other localities, the occurrence of a single specimen is all I have been able to chronicle, while others of greater general rarity have been frequent, and, in addition to the pleasure I have derived from a quiet acquaintance with their habits, I have been fortunate enough to record a few interesting facts respecting them.

The former part of this remark applies to the first species in the order—the Great grey Shrike (*Lanius excubitor*)—a single male bird being the only one I have met with here; this was perched on a hedge in Thoresby Park, from which it flew on my approach. I am aware of one or two specimens having been seen by others, and I have seen one that was shot in Nottingham Forest—a male in fine plumage—but it is here only a straggler. Its congener, the Red-backed Shrike (*L. collurio*) is, on the contrary, a regular visitor, and I have repeatedly observed it on the hedges bordering plantations, which appear to be a favourite place of resort. In these hedges its nest is usually placed, and they offer advantages also for the capture of its prey. It is a noisy bird during the breeding season, its clamorous cries of defiance on the approach of danger only exposing it to greater notice. Amongst its eggs which I have taken I have one pretty specimen, the ground of which is suffused with a pale pink, and the encircling band of spots at the larger end is a deep red.

It has long been asserted that the shrikes fix insects on thorns in order to decoy small birds within their reach. This, though disputed by some, is believed in by others, and I confess myself of the latter number.

Rennie relates, in his Architecture of Birds, that a friend of his expressing his doubts of this habit, he undertook for his own satisfaction, as well as his friend's, to endeavour to ascertain the fact, and he soon found within five miles of Lee, in Kent, half a dozen nests of each species. "We discovered," he says, "that near those nests large insects, such as humble-bees, and the unfledged nestlings of small birds were frequently seen stuck upon thorns. We did not succeed in seeing the birds actually impaling their victims, but we ascertained what we considered good proof of the fact; for the peasants, who had never heard of the story," which he says was first promulgated by Heckwelder, "all concurred in affirming that the butcher-birds fix their prey upon thorns, not, however, according to their belief, to allure larger game, but to kill or secure what has been already captured."

The fact of the shrikes impaling on thorns the bodies of small birds which they have killed has been too often observed and too well authenticated to admit of doubt; but I should imagine that the mangled body of a hedge-sparrow or yellow bunting would offer little that was *attractive* to their fellows. With insects I think the contrary is the case; the bird has no need to impale a moth or a humble-bee in order to devour it; and it is my belief that this is practised as a lure to entice small birds. This idea is strengthened by the careful manner in which the insect is generally transfixed. I was one day rambling on the skirts of a wood which divides Thoresby Park from the forest, where, amongst the heath and fern surrounding it, many young thorn trees were growing, the hedge inclosing the wood being a favourite haunt of the red-backed shrike. Something on one of these thorn-bushes caught my eye, and on going up I found it was a large egger moth impaled on a strong thorn. The thorn projected at least half an inch through the body of the moth, and so little was it injured that it was difficult to understand how the bird placed it there. But the bush stood in such a secluded spot, surrounded by ferns as high as my breast, and at a distance from any path, that I could come to no other conclusion than that it was the work of the shrike. If it had not been intended as a decoy, such a *bonne bouche* would hardly have remained undevoured by the shrike, for the body was quite dry, and it had evidently been there some days. However, the specimen was such a fine one, and so artistically "pinned," that I cut off the branch as it was, and carried it home.

A much rarer member of this family is the Woodchat Shrike (*L. rufus*), of which I am glad to record the occurrence, one having been killed in the forest near the entrance to Thoresby Park called the Buckgates, in May, 1859, by Mr. H. Wells. It is abundant on the Continent, and the late Mr. Hay described it as breeding in the Netherlands. Its eggs have the same zone of spots as the other species, but differ somewhat in the ground colour, which is in some cases tinted with blue.

I am able to include both the pied and spotted flycatchers in my list. The former (*Musicapa luctuosa*) I had the pleasure of first seeing at the end of April, 1854, though only for a few minutes. I imagine from the early period of its appearance that it must have been merely *en route* to those more northern counties which appear to be its strongholds. It was perched on a projecting stake in a hedge by the roadside, a little way out of the village, and it immediately attracted my attention by its active movements as well as by its rarity. The morning was genial and sunny, and the bird, a male, was catching flies most vigorously, as if it had not previously broken its fast, springing from its perch into the air and again returning, at the same time uttering in a very decisive manner a short note resembling " chuck." For a few minutes I watched its motions with great interest, when it became alarmed at my presence and flew away. I should not be surprised to find this species breeding with us; we have many spots which are exactly suited to its habits, and it is found in both the adjoining counties of Derbyshire and Yorkshire; but at present I only know it as a visitor.

I have since met with the Pied Flycatcher several times in the forest, and I hope my anticipations as to its breeding may be realized. Mr. Wells of Cockglode has killed several of them in Birkland.

The Spotted Flycatcher (*M. grisola*) is very plentiful, but is comparatively rarely seen in the forest until after its young brood are able to fly. The gardens around the village and the meadows along the margin of the stream are its principal resorts, the latter especially so. The

E

banks are fringed with many old and decaying pollard willows, and in some of their cavities it delights to place its nest. Often on a quiet summer's evening, when strolling alone down the river's side with my fishing-rod, have I remarked it in such situations, and while I have been silently making a cast with my flies, intent on hooking a trout, it has been equally intent on making its captures, and seemed to pay little regard to my presence. Now, with a short spring, it would seize some gnat from the cloud dancing in the air, and immediately return to its perch; now, making a longer excursion after a Mayfly as it hovered over the stream, or with more devious flight pursuing a vagrant white butterfly. Often on such occasions, when the setting sun tinged everything with gold, and the peaceful calm was scarcely broken by the murmur of the rippling water, have I watched this bird, and have been much interested by its busy quietness, which at times seemed to change to one of listlessness or melancholy—an idea which was strengthened by its plaintive chirp.

Its stay with us is briefer than that of most of our summer visitors, and though I have seen it as early as the first week in May, it seldom makes its appearance until towards the end of that month, and leaves us about the close of September. Its nest is a careless structure; those I have found in the willows were usually made of grass, with sometimes a few slender twigs at the base.

As settled residents or visitors we number many members of the thrush family, and it was with no less pleasure than surprise that I first made the discovery that the Dipper (*Cinclus aquaticus*) was to be found in our little trout stream. I once, whilst fishing, saw at a distance a bird which I failed at the time to identify, and a boy who sometimes brought me eggs, described a bird which he had seen in the stream, which (as I never expected to meet with the dipper) I tried in vain to make out, and felt sure he must be mistaken. Further research, however, cleared up my difficulties, and the sight one summer's evening of a veritable dipper disporting itself in the water settled the matter, much to my satisfaction. I must add that I have never been fortunate enough to meet with it again, though they have been seen once or twice by others. I have in vain sought for the nest, and from what I have read of its usual haunts we have few suitable spots for its erection; but the stretch of stream in which it was seen is very solitary, the channel being confined on one side with high sandstone banks covered with furze, while numerous plantations and ash-holts border the other, the whole bearing the name of New England.

Quiet inoffensive bird as the dipper is, yet it has been lately accused of feeding on the eggs of the salmon and trout. I believe this charge, like some which are brought against others of our feathered friends, cannot be substantiated, but is nothing more than a hasty conclusion. The dipper doubtless frequents the streams where salmon and trout spawn, but it is to feed upon one of the worst enemies of salmon ova—the larva of the Mayfly, as well as water-beetles and molluscs. Let those who believe in the poor little dipper's crime, open the stomach of the next they kill, and I am much mistaken if they will not coincide with me in a verdict of " not guilty."

The Missel Thrush (*Turdus viscivorus*) is very common, and in the winter frequents the forest and parks in large flocks. During this season they fly in a scattered, loose order, in a series of undulations, and seldom for any great distance at a time, sometimes alighting on the ground and searching for food, at others perching on the tops of the highest trees. I have found them on these occasions very wary and difficult to approach, one of the party apparently acting as sentinel, and on the alarm—a loud harsh note—being given, the flock take wing one after the other, and never, as far as I have observed, simultaneously.

After they have paired, orchards and gardens are their favourite breeding places. A large pear tree in a garden near my own was chosen several years together, evidently by the same pair, as the site for their nest; it was placed on a large bough, close to the trunk, and was by no means so carefully concealed as is usually the case with this species. I have often been amused by their pugnacity, especially after the eggs were hatched, any bird that came near their domicile being instantly chased away. The pleasure grounds at Thoresby are much frequented by them, and I have found their nests on oak trees in the midst of the forest; one that I took on May 11, 1853, was placed in the fork of an oak, about twenty feet from the ground.

Mr. Rennie is rather severe on writers who, in describing the nest of the missel thrush, have omitted to mention that mud or clay is used in its construction. He and many other naturalists describe the skeleton of the nest as being composed of twigs, incorporated with clay; while, on the other hand, authors of equal observation and veracity leave out the mud or clay altogether. I am inclined to believe that it is much as with the travellers who disputed about the colour of the chameleon, "both are right, and both are wrong." My opinion is based upon the simple fact, that the nest I mentioned above as having been found in the fork of an oak, had no clay or mud whatever in its structure; for having pulled it to pieces I found it was made of birch twigs, with moss worked into the interstices, and lined with a quantity of dry grass; the outside was neatly ornamented with bits of lichen like that on the boughs on which it rested, withered oak leaves being here and there inserted and left to hang down and add to the general concealment. In other nests I have found clay used, but it is evident that their method of construction is variable.

In 1853 a nest came under my notice which, from the singularity of its lining material, proves what I have said, that the missel thrush does not adhere to one unvarying rule in its constructive plans. On the 5th of May in that year one of the labourers in the pleasure grounds at Thoresby Park, observing something white hanging from the bough of a cedar which stands in the grounds near the entrance from the courtyard, climbed up the tree, and at the height of about thirty feet from the ground discovered the nest of one of these birds, daintily lined with a piece of lace, and a lace cap and collar. These articles had been missed, and nothing could be learnt respecting them until the white string of the cap, dangling in the wind, led to the discovery of the lost treasures. They were carefully laid round the interior of the nest, and on the unwonted lining, which was not in the least soiled, were deposited two eggs. The drying ground is situated on the outskirts of the pleasure grounds, about forty or fifty yards from the tree, and from this spot had the articles been purloined by the mysterious thief.

The missel thrush is an early singer, and though it

does not possess the mellow tones and variety of modulation of the song thrush or blackbird, yet it is no mean songster, and much more attractive than some authors describe it. I have heard the woods ring again with its music, the height at which it perches while singing (generally on the extreme point of a tall larch) effectually subduing the loudness of its song, which is much greater than that of either of the two I have mentioned. The absence of other songsters during the months when it is most in song makes its music the more welcome. I have heard it frequently about the 19th or 20th of January, sometimes when the day was fine and sunny, at others when it was cold and stormy, as befits its best-known name of the "storm cock." I must confess that to me its song ringing at such times has a very charming effect; and, instead of its "loud, untuneful voice," as Mr. Knapp calls it, "being like that of an enchanter calling up a gale," it has seemed to me to herald forth with gleesome heart the approach of the more genial days of spring. The reputed favourite food of this thrush, the berries of the mistletoe, is most abundant in the district, growing chiefly on the whitethorn. I have no doubt that the missel thrush assists greatly in the propagation of this curious parasite; I used to think that the idea of the seeds germinating after passing through its stomach a mistaken one, for I conceived that the action of the gizzard and stomach would effectually destroy all their vitality, but in this I must confess myself mistaken. Its agency as a disseminator of the plant is exercised also in another way. The berries are exceedingly viscid, and the seeds frequently cling tenaciously to the bill of the bird, who, to rid itself of them, is compelled to rub its bill on the bough of a tree, and thus the seeds are unwittingly placed in the best position for germination in the clefts and crevices of the bark.

The Fieldfare (*T. pilaris*) is as abundant with us in the winter season as it is in every other part of the kingdom. The forest and parks are more frequented by them than the fields, the former being in many parts thickly studded with hawthorn trees, while under the modern system of farming, the hedges surrounding the fields are rarely allowed to grow to the height at which they usually flower, but are kept low and well trimmed. These hawthorns are of great size and age, and rarely fail to bear an abundant crop of berries; indeed, in the spring they are so profusely covered with bloom as to appear at a distance like huge snowballs; while in autumn, when the berries are ripe, they are masses of scarlet. Of course such abundant provision is duly appreciated, and attracts large flocks of fieldfares and other birds; I have also seen them feeding on the berries of the mistletoe. When these fail, and when long-continued frost or snow cuts off their insect food, they are put to great straits, their usually wild and wary character is exchanged for a bold and fearless one. At these times they associate with the sheep in the turnip fields, and frequent the clumps and plantations of beech trees in Thoresby Park and the forest, where they scratch through the snow into the heaps of withered leaves underneath, to search for beech-nuts, of which they are particularly fond.

The earliest day on which I have noted their arrival has been the 9th of October; they generally leave us about the end of April, seldom at that time appearing in flocks, but in scattered pairs. In 1847 I met with several pairs on the 9th of June.

They most frequently roost on the ground, but I have sometimes aroused them from the old oaks.

The redwing (*T. iliacus*) visits us in greater numbers than the fieldfare, and, both species feeding on the same food, are often seen mingled together. They both frequent the same localities, and whilst reaping the benefit of an abundant supply of food, are equal sufferers during hard weather; indeed, I think they feel the effects of scarcity even sooner, and become more quickly emaciated, some I have shot in a severe season being mere skin and bone. At such times they are more fearless and familiar than the fieldfare, and I have approached within three yards of some when searching amongst fallen leaves for beech-nuts, and this close to Thoresby House. Even then they manifested no alarm, not even taking wing, but merely hopping a few yards further away, and busily continuing their scrutiny. In the same winter (1849) I disturbed some out of the garden attached to my house in the village; they flew away on my approach, settling in some nut trees a few yards off, and uttering at the same time a plaintive cry. In open weather I have found them more timid than the fieldfare, and they perch more in trees than the latter generally do.

The Song Thrush (*T. musicus*) is a very abundant species; in some of the plantations you may meet with a nest every few yards. Like the missel thrush it is an early singer, and I have remarked as soon as the 1st of February, many perched on the top of furze-bushes in the coverts filling the air with music; and even at that season they will continue their song from morning to evening. Though generally commencing to build in March, instances of earlier incubation are not wanting;

and I have a record in my notebook of finding a thrush sitting on five eggs on Feb. 22, 1859. The season was particularly mild and open, and many other birds were equally early in their nesting. A singular occupation of a thrush's nest by other species occurred a few miles from us in 1846. A pair of thrushes built their nest in some ivy on the mansion of Mr. Simpson of Babworth, and notwithstanding the spot they had chosen was close to the door, and that persons were constantly passing, they reared in safety four young ones, who duly took their flight. Immediately on their departure a pair of blackbirds took possession, and after effecting a few repairs the female laid her eggs, which she hatched, and the young safely left their birthplace. No sooner was this done than a couple of spotted flycatchers became the tenants of the domicile, but not contented with its condition they built one of their own nests within it; five eggs were laid and hatched, and thus a third brood was successfully reared in the same nest.

Though such attractive denizens of our woods, they are sadly destructive to the fruit in our gardens, and in an hour will strip a large currant bush of the whole of its crop; indeed, during one summer they, in conjunction with blackbirds, frequented my garden in such numbers as to make serious inroads on my fruit. Defensive measures were in vain; nothing in the shape of whirligigs or scarecrows had any effect, for their numbers did not appear to be at all diminished; "the cry was still, they come," and only when the fruit was gathered and devoured together did they cease their visits. I believe half my crop of gooseberries and currants, which that year was unusually heavy, was eaten by them.

The eggs of the thrush are sometimes found of a uniform blue, without mark or spot. I have frequently taken one of these plain ones in a nest where all the rest were of the usual spotted kind, and on one occasion I found a nest in the Lawn plantation in Thoresby Park containing five eggs, four of which were of this spotless variety, while the fifth had five or six distinct spots of black towards the larger end.

Albino varieties of the thrush are not uncommon; one, two, and even three white ones have been taken out of the nest where the plumage of all the rest had the usual markings, and one instance is related by the Rev. F. Morris of a nest being found by Dr. Moses of Appleby, where all the five birds were albinos, with scarlet eyes.

The idea has occurred to me, although I have had no opportunity of attempting its verification by experiment, whether these white varieties may not be the produce of the plain blue eggs. At the same time I am bound to say that, though I have repeatedly taken eggs without a spot, I have never met with more than one bird whose plumage differed from the usual markings. This was not pure white, but had a large number of white feathers mingled irregularly amongst the brown ones, chiefly in the back and wings.

The Blackbird (*T. merula*) is as abundant a species as the thrush. The rich black plumage of the adult male, so compact and glossy, relieved by the deep orange of the bill and eyelids, makes it an attractive and favourite bird, its bright eye harmonizing with the vivacity of its movements. It is a very wary bird, and when disturbed in gardens (which are much resorted to) it is unwilling to take wing as long as there is any covert, beneath the friendly shade of which it bounds, rather than hops, rapidly and silently away; but when forced to leave its shelter it flies off hurriedly, with vociferous notes of alarm, though seldom to any great distance. During the winter of 1860-61, the cold was so intense that even the blackbird lost his wariness, and a fine male repeatedly came for food to my kitchen door.

In gardens, as I have before mentioned, it is equally destructive to fruit as the thrush; but I remarked that nearly all the birds of this species which I saw in the summer I have named were females, or young birds of the year.

Their staple food consists chiefly of berries of various kinds, but I have seen them devour earthworms with great avidity, and snails, too, are a favourite repast for them, as with the thrush. At all times of the year, but especially during the winter months, I have often watched them at the bottom of the hedgerows, breaking the snail-shells with repeated blows of their bills, or sometimes by dashing them on a stone, and have been surprised to see the quantity of broken shells they have left in one spot. A writer in *Chambers' Edinburgh Journal*, quoted in Sweet's Warblers, mentions an instance where a grassplot was quite furrowed and disfigured by a number of blackbirds, who were found on examination to be feeding on the larvæ of the cockchafer with which the ground was infested.

The blackbird has usually two broods in a season, and sometimes more, generally in the same nest, though occasionally a fresh one is built. I have known an instance where a pair brought up three broods in one season, and in the same nest; the first brood consisting of five, and the second and third of three each.

The eggs often vary greatly in shape, and I have one in my collection which scarcely tapers at all, but is almost cylindrical, with obtusely rounded ends. I have also taken some which distinctly differed from the usual markings in having the ground colour of a clearer blue, while they were rather sparingly freckled with rich reddish-brown spots, with here and there a more prominent dash, approaching the general appearance of the eggs of the ring ouzel.

With regard to the song of the blackbird, I am half inclined to consider it superior to that of the thrush; though lacking the variety and continuance of the latter, it has a far richer and fuller tone, its few notes being most exquisitely modulated, and to hear it at sunrise pour forth its joyous song, when perched on the extreme top of a high tree, is peculiarly pleasing; it seems to be not merely for the amusement of its mate, but the expression of intense happiness, a hymn of praise to the great Creator.

There are few of our British birds of which we know so little generally as the Ring Ouzel (*T. torquatus*); it is by no means what may be called common anywhere, and its partiality for wild and unfrequented localities, such as the mountain and the moor, tends still further to diminish our acquaintance with it. In addition to this, its natural habits are shy and retiring, and excepting when it assembles for its migrations it is only met with in single pairs. In other parts of the country its visits are chiefly confined to the spring and autumn, though occasional instances are recorded of its breeding in places at variance with its natural habits. I was pleased to meet with a pair in the summer of 1856. The nest, which was very similar to a blackbird's, was placed in a bush in the forest about four feet from the ground, and contained three eggs. Their ground colour is pale greenish blue; two of them are speckled and dashed all over with reddish brown, but the other has the markings more closely distributed, and at the larger end they form a confluent zone, having a suffused appearance, and quite concealing the ground colour.

Another pair were seen at the decoy at Houghton on the 6th of April, 1858, but, with the usual fate of rare birds, they were shot, or I have no doubt they would have bred also.

A male in immature plumage, with the crescent on the breast distinctly marked in light brown feathers intermingled with one or two of a still lighter shade, was shot at Edwinstowe on the 26th of November, 1856.

These are all the instances of its occurrence with us that I am aware of.

The numerous family of the warblers is a very attractive one. The sweetness of the song of the majority, their compact and sprightly appearance, and the liveliness of their manners, combine to make them general favourites. The greater part of them are closely associated with the spring and summer seasons, and we are accustomed to look forward to their arrival with great pleasure.

The unpretending dunnock, or Hedge Sparrow (*Accentor modularis*), in its neat and sober-coloured dress, is common, as it is everywhere. It cannot be called gregarious, seldom more than two or three being seen together or near each other. It is always busy, though never in a hurry, and seems an especial pattern of that valuable qualification of minding one's own business. It hops—or rather glides—quietly amongst the twigs of the hedges or the currant bushes; and its chief food being insects, and being, moreover, especially active in the destruction of caterpillars, its labours in the gardens are of no little value. Its low but sweet song is perfectly in keeping with its appearance and habits, and not only is it a summer warbler, but, residing with us all the year round, it " cheers the winter with its melody." I have heard it at the beginning of February singing as gaily as in June. It is a hardy little bird, and an early nester.

How different in disposition is the Robin (*Sylvia rubecula*). Deeply enshrined in household memories is "the little bird with bosom red," and never to be forgotten the childish delight with which we have pored over the sad fate of the lost but loving children, while

> "Robin Redbreast faithfully
> Did cover them with leaves."

He wins our admiration and regard by the frank confidence with which he approaches our dwellings, and the fearlessness with which he takes the offered crumbs, turning up his large dark eye the while. And yet this bird, so familiar and so welcome, is amongst his fellows of a most quarrelsome disposition. I have seen two robins meet to do battle with all the boldness of the game-cock, lowering their heads and setting up their hackles in a similar manner, and leaping at each other with the utmost fury, utterly regardless of my approach.

Other birds are not exempt from its pugnacity. I have often been amused when a party of small birds, sparrows, chaffinches, &c., have been engaged in searching about in the garden or the yard, to see their quiet labours disturbed by the sudden appearance of a robin amongst them. Lowering his wings, he rushed first at one and then another, until he had driven them all away and remained master of the situation, the others not venturing to contest the matter.

But in spite of its pugnacious disposition it will ever hold a chief place in our regard. Like its winter neighbour the dunnock, its principal food is insects, and no corner of the garden is overlooked in its search for those tiny ravagers, which, if left unchecked, would do us so much mischief. In addition to this the robin is one of our most pleasing songsters, and its sweet notes are heard, with few intermissions, all the year round. In the spring there is more liveliness and vivacity in its tones than at other times, though amidst the choral harmony that then prevails it attracts but little notice. But when our summer friends are fled with the fading flowers, and the "sere and yellow leaf" comes whirling from the tree, the robin's song awakens our attention. Yet it lacks the joyousness of spring, and in sympathy with the departing season it seems to breathe a plaintive and melancholy strain, bidding us, as it were, to remember that we "all do fade as a leaf," and turning our thoughts to that haven of rest where nought is touched by "decay's effacing fingers," but, fresh with eternal spring, the redeemed of the Lord shall dwell there for ever, "and there shall be no more death."

Keble quotes from a friend some sweet lines in his Christian Year, which I am tempted to transcribe:—

> "Unheard in summer's flaming ray,
> Pour forth thy notes, sweet singer,
> Wooing the stillness of the autumn day;
> Bid it a moment linger,
> Nor fly
> Too soon from winter's scowling eye.

> "The blackbird's song at eventide,
>   And hers, who gay ascends,
> Filling the heavens far and wide,
>   Are sweet. But none so blends
>   As thine
> With calm decay and peace divine."

It is rather singular that, while the robin remains with us throughout the hardest winters, it suffers much less from cold than many other birds. It is stated by Bechstein that in Germany it migrates to warmer climates at that season, leaving in October and returning in March, the few that venture to remain paying for their want of prudence with their lives. A writer in *The Field* on the natural history of Malta states the same fact. He says: "Robin Redbreast comes hopping in about the same time, and through the sunny winter day sits on the bare bough of the fig, 'piping disconsolate' until early spring, when, with such of its kin as have passed Malta in autumn and struck Africa, penetrating even to the oasis of the Sahara, again it returns to Sicily and Southern Europe to rear its young. Why is the robin of the south such a valetudinarian, whilst his brother of the north braves the severest winter of England?" The answer to this question is not easy to give, but it opens a point of much interest. With us it is very hardy, and perhaps one reason why it bears the cold better than others may be that, from its fearless and familiar nature, it obtains food in places where other birds dare not venture, and thus is better fortified to resist the rigours of winter.

The apex of the gable end of a building is a spot often chosen by the robin from which to pour forth its song in the autumn, and even in the depth of winter; and I once in January, at the close of a long frost, noticed five robins in full song at the same time, and within a stone's throw of each other, three of them occupying the position I have named, while the other two were perched on the corners of high chimneys.

I have met with the robin in the wildest and most solitary parts of the forest, but chiefly during the summer; yet, though favourable as such seclusion is, I have rarely found them nesting in such situations. A sunny bank at the foot of a hedgerow, or sometimes in the lower part of the hedge itself, and various positions in gardens, are most commonly chosen. I have several times found a nest placed on the top rail of a row of espalier white currants in my own garden, about four feet from the ground.

The eggs frequently vary in their markings, but I have never found the eggs in the same nest to do so. I have seen them where the ground colour has only been removed from white by the faintest possible tinge, and they have been uniformly marked all over with pale red. Generally, however, the ground is of a delicate pale reddish-brown colour, freckled with a slightly darker shade, and forming at the large end a dark and distinct zone.

The pretty Redstart (*S. Phœnicurus*) is one of our handsomest summer visitors, and though abundantly distributed, its quiet and wary habits would lead a careless observer to think it uncommon. But this is far from being the case. If you ramble in the forest, following one of the grassy paths which wind their way amongst the ancient oaks, or seat yourself on a moss-covered root, you can hardly fail to see this lively little bird restlessly flitting about. Its nest is most commonly placed in a decayed oak-tree, whose hollow trunk and

F

branches afford numerous safe and sheltered sites. Sometimes so small a cavity is chosen that its nest is perfectly safe from the usual poachers of eggs, including even the daring schoolboy; at others it is fixed inside the hollow trunk, supported by some rugged projection. Amongst the branches of these trees, or those of the hawthorns which are scattered about, they may be seen searching for caterpillars and other insects, and now and then darting on outstretched wing after a gnat or beetle. Sometimes they may be observed busily engaged on an anthill, both ants and their eggs being a very favourite food; but they soon retire from notice unless the spectator is very still and quiet.

Amongst the oaks in Birkland it is abundant, and especially so in an oak plantation on Budby South Forest, which is divided from the open forest on the southern side by a high hawthorn hedge, outside of which runs a broad green drive for more than half a mile. This is a very favourite walk, and many an hour's enjoyment I have had in watching the redstarts here. During the early part of the summer the males alone are visible, and on the approach of any one they utter a quick impatient note, flitting restlessly about, now perching on a projecting twig of the hedge, now on the top of a furze bush, flirting the wings and tail with a quick motion, and showing conspicuously the bright colour of the underside of the latter and its coverts, from which they derive their best-known name of the "firetail." Now they will fly off to one of the upper branches of an oak, their voice and manner expressing their dissatisfaction at your presence; but later in the season—when the young are hatched and able to cater for themselves— they do not manifest their hostility to intruders, and if such walk quietly and slowly along, the redstarts will take no further notice than to flit a few yards further on and continue their occupations.

In this neighbourhood they rarely frequent gardens, though I have occasionally met with them there ; but in every instance, though approaching within a few yards of the house when no one was visible, they still retained all their usual shyness, and instantly flew off when any one came in sight. I never met with their nests in such situations.

I always admire the eggs of the redstart ; their colour and shape are both peculiarly elegant, and when lying in their nest in a cavity of an old tree, they form a pleasing picture, their tint harmonizing while strongly contrasting with the rich brown of the decayed wood around them.

Those of my readers who possess the spirit of a naturalist, can easily imagine the pleasure which arises from a first acquaintance with a rare species which they have only read of or seen in a museum. Such will sympathize with me in my delight and surprise when I first met with the Black Redstart (*S. tithys*), and found it not only a visitor but actually breeding with us. My first acquaintance with it was the discovery, on May 17, 1854, of a nest in a thorn hedge by the side of the road leading from Ollerton to Edwinstowe. It was placed about four feet and a half from the ground, and was constructed of dry bents, intermingled with a little moss, and lined with hair. When I found it, it contained four eggs; had it remained undisturbed, I have no doubt they would have been increased to the usual number of six, as the female was on the nest. As it was, I appropriated them as a valuable addition to my collection. This, however, was not a solitary instance, for two years later, on May 13, 1856, another nest was taken from the same hedge, near the place from which I had taken the previous one; it contained one egg, which was brought by the finder to me. A third nest was taken the next day at Ollerton; it was placed in the side of a cattle hovel, amongst the thorns with which the upright framework was interlaced, and was constructed of dry grass only, and lined, as were the others, with hair. The second nest had moss mixed with the grass, like the first.

My satisfaction at the discovery, so far as I am aware, of the first instance of the blackstart breeding in England (for I perceive Mr. Newman does not include it as so doing in his recent Zoologist List), was all the greater from their being no possibility of the fact being contested, as in addition to my seeing the bird, its eggs cannot be mistaken for any others. Without possessing the least polish, they nevertheless have a very peculiar gloss, and are of the purest white, with an extremely delicate semi-transparent appearance, quite unlike those of any other British bird.

From the circumstance of the two first-named nests being placed in the same hedge, I should infer that they were the work of the same pair of birds, and this probability is of course increased in the case of a rare species.

It is singular that their nests were placed in situations so different from those authors describe as usually frequented by them. Bechstein, to whom the bird was well known, says:—"They build in rocks and holes of walls, but especially in lofty old buildings, on timbers of roofs where the nest can stand alone on a beam without support." No similarity to its native haunts could have tempted it to remain with us, for we have nothing in the shape of rocks for many miles around us, and the fact seems to be one of those inexplicable ones which baffle all our conjectures to find a cause. I have not met with it since, but I should not be surprised to find a few pairs breeding regularly in the neighbourhood.

Those pretty little birds, the Stonechat (*S. rubicola*) and the Whinchat (*S. rubetra*), are very common, particularly on the furze-clad parts of the forest, which they much enliven by their active and restless habits. They are almost exclusively found in wild localities like these, seldom intruding on the limits of cultivation. The stonechat resides with us all the year round, but I think not in equal abundance, a partial emigration appearing to take place in the autumn, while their numbers are increased in the spring. It is a snug-looking, compactly-built bird; and the male in the breeding plumage, with the deep black head and rufous breast, is really handsome, though there is much variation in the distinctness and brilliancy of the colours. Perched on the top of a furze bush or a prominent sprig of heath, these birds utter their singular notes, "chat, chat, chat," from which they take their name; and though rather shy and wary, I have often called them close to me by rapping two stones together, and thus producing an exact imitation of their call. While perched in this manner, they jerk their tail and wings simultaneously with the utterance of their cry, then perchance dart to the ground to capture an insect, and again flit to their post of observation, rarely remaining many minutes in one spot. They seem to be very constant in their attachments, seldom being seen otherwise than in pairs even during the winter; and I think that only the old birds remain during that season, the young of the year appearing to migrate to some other locality.

The whinchat is a summer visitor with us, as elsewhere, but during the few months of its residence it is more abundantly distributed than the stonechat. It frequents the same localities, but is not so exclusively confined to the open forest or moorland; I have often met with it in pasture fields. Its habits are almost precisely those of the stonechat, though it shows more fearlessness, perching on the bushes of furze or heath, or sometimes on the hedges by the roadside. Its ordinary call is scarcely so *stony* in sound as the former, whilst its song is more musical; and I have heard it singing very sweetly while it hovered over a bush before perching, its notes much resembling those of the skylark.

The nests of both species are placed in similar situations at the foot of a bush of furze, or gorse, as it is called locally; they are very difficult to find, and when discovered are by no means easily obtained from the midst of the natural *chevaux de frise* that surround them. The difficulty of finding that of the whinchat is greatly increased by the covered entrance leading to it, and I have often searched in vain when I felt sure, from the presence of the birds, that the nest was close at hand.

That graceful and chastely-coloured bird, the Wheatear (*S. Œnanthe*), is a regular summer visitor, but is confined to two or three spots—viz., Oxton Warren and Boughton Brake (both of them being rabbit warrens sparsely covered with furze bushes), and occasionally on that part of the forest adjacent to the toll-bar on the outskirts of the village.

Though always to be met with in the two first named tracts, it is never seen except in pairs, and these but thinly distributed, though most frequent on the warren at Oxton. This, traversed by the road leading from Ollerton to Nottingham, is a lonely, quiet spot, and I have watched there with much interest these pretty but wary birds. From being accustomed to see persons passing along the road, they are not quite so shy as usual, and will allow any one to approach within a few yards before taking flight. Perched on a grass-covered molehill or a large stone or clod, they seem always on the watch, turning the head quickly from side to side, while the body is carried very erect. They hop rapidly, but their flight is low, and is maintained for very short distances. Their movements while catching flies and other insects are very lively; but they are continually occupying for a few moments some little prominence, and again flitting about after their food. I have once or twice seen them perch on the top of the walls of turf that surround the warren, but I never knew them to do so on either bush or tree. The wheatear is comparatively a silent bird, its faint warble being seldom heard, and its call-note, which it utters while hopping or rather running about, consisting of a single "chat."

The Grasshopper Warbler ($S.\ locustella$) is more plentiful than it appears to be. It is so fond of concealment, and so shy and watchful in its habits, that even in the places which you know it frequents it is difficult to catch a glimpse of it; and whenever this is accomplished, it seems, from the way in which it creeps or rather glides through a bush or hedge, as Mr. Yarrell justly remarks, "more like a mouse than a bird." I have seen it so repeatedly in the furze and underwood in some parts of the forest, heard its stridulous note so constantly, and, in addition, have obtained the bird itself, that I am perfectly satisfied it is not by any means rare; and yet I have never been successful enough to discover the nest. Many times have I watched the birds with the utmost caution in the hope of tracing the female in her approaches to her eggs, but the habit of forming a covered passage has increased the difficulty, and I have always been disappointed in my endeavours. They generally arrive about the latter end of April—but I have known a specimen obtained on the 13th of that month—and leave us again at the beginning of September.

Amongst the common birds, the Sedge Warbler (*S. Phragmitis*) is conspicuous, and its fearless and garrulous habits render it very amusing. That which would frighten away another bird only has the effect of exciting it to louder outbreaks of noisy mirth, sometimes uttered as though in defiance of all comers. It chiefly frequents the neighbourhood of our various small streams, though I have known its nest placed at some distance from water. The bottom of a hedge bordering a stream is a favourite position, where the long grass and weeds have formed a friendly screen. I have seen a nest in a very different place—viz., on the head of a pollard willow, several large tufts of grass which had there taken root in the partially decayed wood, effectually concealing it from the observation of any one on the ground; and I, perchance, should not have discovered it had I not one day, when fishing, climbed the tree to free my artificial flies from a bough in which they had hooked.

The song of the sedge warbler is a kind of medley,

and seems as if composed of imitations of various birds. I cannot, however, agree with those who consider this species and the whitethroat as mocking birds. Although undoubtedly the notes of the skylark and the swallow may easily be recognised, and especially those of the house sparrow, yet why should it be supposed that these are merely imitations? Such an idea has not much show of reason in it, for the house sparrow is one of the last birds whose note is likely to be heard by the sedge warbler. I have met with a passage in Mr. Rennie's Habits of Birds, which I think places the matter in so clear a light that I am tempted to quote it. He says: " Amongst some hundreds of these birds which we have listened to in the most varied situations in the three kingdoms, all seemed to have very nearly the same notes, repeated in the same order; a fact which appears to us to be fatal to the inference of the notes being derived not from one, but a number of other birds. For if this were so it is not possible that these imitated notes should all follow exactly, or very nearly, the same order in the song of each individual imitator in different and distant parts of the country. The close similarity of the notes to those alleged to be imitated cannot be denied, but, taking all the circumstances into account, we think it much more probable that these resembling notes are original to the sedge bird, and that we might with equal justice accuse the swallow and the skylark of borrowing from it."

The above exactly corresponds with my own experience. I have heard the same series of notes continually occur, and this repetition of the strain has been always rendered the more noticeable by the harsh chirrup of the house sparrow occurring at intervals which were as regular as the song of any other of our warblers.

The Reed Warbler (*S. arundinacea*) I have not met with nearer than Nottingham, where it is tolerably abundant in the reed beds on the banks of the Trent.

I think there are none of our warblers that are so truly local as that queen of them all, the Nightingale (*S. luscinia*), although it is mysterious how exactly the line of demarcation is drawn between one district and another, where no apparent difference exists which can account for their frequency in the one and their absence in the other. It is a regular visitor in some of our woods, though not by any means numerous. Ollerton-corner Wood is the most favourite spot. There I have heard it make the forest echo with its melody, and on still evenings even as far as my own house, a distance of half a mile. One summer a large poplar in the garden of a farmhouse in the village was the constant resort of a male bird in the evening. His station was usually near the top, and here for an hour or two at a time would he pour forth his song, taking no heed of the passers-by, who continually stopped to listen. I was not able to ascertain whether his mate had her nest in the vicinity, for I never saw her; but she most probably had.

Their arrival in the forest is generally during the first week in May, the second of that month being the earliest day on which I have heard their song. I have seen a male and female on the 7th of May busily searching an anthill in one of the grassy rides bordering the wood I have named above. Their motions were quick and full of vigour, but on perceiving me they flew up into one of the oaks, from which they quietly watched me for a few minutes.

The question has often been discussed as to whether the song of the nightingale is merry or melancholy, and many are the authorities both in poetry and prose who have been ranged on either side of the controversy. I do not presume to decide the matter, or to set aside the verdict of the many well-qualified judges who have expressed themselves on this *quæstio vexata;* at the same time, as a close observer, I must reserve to myself the right to differ. My own opinion is, that though it lacks the ringing hilarity of the song thrush, I should never call it melancholy. Coleridge's beautiful lines exactly embody my own thoughts.

> "A melancholy bird? Oh! idle thought—
> In Nature there is nothing melancholy.
> But some night-wandering man, whose heart was pierced
> With the remembrance of a grievous wrong,
> Or slow distemper, or neglected love
> (And so, poor wretch! filled all things with himself,
> And made all gentle sounds tell back the tale
> Of his own sorrow), he, and such as he,
> First named these notes a melancholy strain,
> And many a poet echoes the conceit.
>   We have learnt
> A different lore; we may not thus profane
> Nature's sweet voices, always full of love
> And joyance! 'Tis the merry nightingale
> That crowds, and hurries, and precipitates
> With fast thick warble his delicious notes,
> As he were fearful that an April night
> Would be too short for him to utter forth
> His love chant, and disburden his full soul
> Of all his music! * * * *
>   * * * * * Far and near,
> In wood and thicket, over the wide grove,
> They answer and provoke each other's songs,

> With skirmish and capricious passagings,
> And murmurs musical, and swift jug, jug,
> And one low piping sound more sweet than all,
> Stirring the air with such a harmony,
> That should you close your eyes you might almost
> Forget it was not day."

There is nothing sad or sorrowful in its sweet tones; but in perfect harmony with the quietude of a summer's evening, when all the toil and bustle of the day is hushed, it breathes a sense of calm and peaceful happiness.

Much, no doubt, as Coleridge has so well expressed in the lines I have quoted above, must be allowed for the state of the listener's feelings. Where the mind of such a one is filled with sorrow or care, he would very naturally invest the song with a plaintive, or even a melancholy character; but, with a mind at rest, and filled with thoughts of Him whose power and goodness have so greatly contributed to our earthly enjoyment, surely it speaks of nothing but thankful gladness—a tribute of praise to the great Creator.

The conjecture above expressed is illustrated by an interesting incident related by the Duke de Cabellino, one of the noble band of Neapolitan patriots who, in 1859, sought a refuge on our shores from the cruel tyranny of the Bourbons. In a letter he wrote on landing in Ireland to the *Cork Daily Reporter*, giving an account of the sufferings which he and Baron Poerio and others endured in the stifling prison cells of Monte Fiesco, he says:—

"A nightingale, as if on a mission from Nature, apparently feeling for our sorrows and solicitude, used to come to the boughs of a mulberry-tree, and with his plaintive song he expressed our griefs, so that he became our friend—the very friend of our hearts. We used to throng to the prison bars to listen and to treasure his loving plaint. Ah! fond fool! he and his tender ditty awakened suspicions amongst the police that we had communicated with the outer world—a blessing, indeed, which they trusted had ended for us. They shouted with their voices and hurled sticks, but in the evening the little nightingale came again and again with his song of solace to us; but his sympathy for patriotism brought his doom—he was shot!"

As a songster, the Blackcap (*S. atricapilla*) is, I think, only second to the preceding species, and well deserves its name of mock nightingale. There is an inexpressible charm about its song, which is wildly sweet and very varied, partaking of the notes of the nightingale, thrush, blackbird, and garden warbler. It pours forth a flood of rich melody, not confined to any set song, but giving a play to his fancy, like the minstrel's fingers wandering amidst his harp-strings,—

" In varying cadence, soft or strong,
He swept the sounding chords along."

This species frequents the gardens along the side of the stream in the village where I have found its nest; but it is not confined to such localities, for I have met with it in the wildest parts of the forest, where the dwarf hawthorns are favourite stations of the male from which to pour forth his song. I have seldom seen more than a pair inhabiting one spot. The pleasure grounds at Thoresby are much frequented by them.

These grounds and the adjoining shrubberies are also resorted to by the Garden Warbler (*S. hortensis*), whose sweet and flute-like song is scarcely inferior to the blackcap; indeed, I think it excels it in fulness and richness of tone, and is certainly more sustained. It is more abundant in this delightful and secluded spot than anywhere I know, and I have frequently found its nest placed in some of the low bushes under the trees at the lower end of the lake. It is usually formed of goose-grass, mixed with small roots, and lined thinly with hair, and sometimes with a little wool. I took a nest of this species which was built of fine goose-grass and slender fibres of a uniform thickness, looking exactly like black and tarnished brass wire, and the singularity of the appearance was increased by the lining of long black horsehairs, which, as well as the materials forming the body of the nest, were laid in concentric circles with hardly any interlacing; the whole formed a rather loose yet neat structure. It was placed in a small box-tree about three feet from the ground, and contained four eggs. The kitchen gardens at Thoresby are also frequented by the garden warbler, where they are very partial to strawberries and raspberries.

The Whitethroat (*S. cinerea*) is one of our commonest summer visitors, and its loud and lively song is constantly heard in our hedges and gardens. It is always amusing, for it seems as if it was ever in a hurry to get through its varied song. This is frequently interrupted, like that of the sedge warbler, by an exact imitation of the chirp of the house sparrow, and it was especially remarkable in a pair that built in my own garden for several years together, and which, from frequenting the same spot for the erection of their nest, I judged to be the same pair. A nut-tree was the place always chosen by the male bird from which to pour forth his song, and from a seat underneath I could watch him without being perceived. His body was in incessant motion, the wings and tail being shaken, the crest prominently raised, and his whole appearance being one of excitement. Most generally he sang while perched on a spray near the top of the tree, but sometimes he would spring up in a singular way as if unsuccessfully trying to balance himself, and then he would hover a few feet above the tree, slowly descending to his perch, and all the time singing with the utmost rapidity. While thus engaged he exhibited very little fear, but would allow me to approach closely before he moved; but he evidently did so unwillingly, not liking to forsake his charge, as the nest was placed in a clump of herbaceous plants at the foot of the tree. The presence of a cat in the garden was always met with loud cries of alarm, and I have seen one that belonged to me, who was a noted birdcatcher, greatly annoyed at these unwelcome attentions, and even shrinking with evident fear from the vigorous attacks which parental love led the little fellow to make on pussy, and in which he seemed quite regardless of his own safety.

The Lesser Whitethroat (*S. curruca*) is almost as common as the preceding species, and is very frequent in the gardens and in the hedges of the neighbouring meadows. My own garden was seldom without a pair during the summer, the clear white of the throat and breast making them very prominent. It does not show itself so openly as the whitethroat, and utters its song while flitting about in the concealment of the bush or tree. In one corner of my garden grew a clump of nut and plum trees, overshadowing an arbour, the sides of which were clothed with honeysuckle which climbed upwards, clasping the boughs of the trees above. The clump was the constant resort of this little bird, and while quietly seated on the bench below, and hidden from its sight by the broad leaves of the nut-trees, I have watched it through the interstices of the foliage with great interest. Here I heard to advantage the low inward warbling, which is not noticed without being close to it, but which is very sweet. This would be interrupted, or rather ended, by the loud, shrill, well-known notes which Bechstein describes by the words "Klap, klap;" while at intervals it would utter several times in succession a hissing kind of note resembling the word "tzee," repeated three times. Sometimes the nest would be placed well hidden in a currant-tree fence close by, and sometimes in a thick privet hedge which shut in my garden from the stream flowing past it. I never could perceive any difference between the male and the female, though I believe the latter is generally described as somewhat paler in colour.

The Wood Wren (*S. sylvicola*), or, as it is better known with us by the name of the "yellow willow wren," regularly visits us, but I have only occasionally succeeded in finding the nest, which has always been placed on the ground, and well concealed with withered leaves. The eggs are rather larger than those of *S. trochilus*, and differ so greatly in the colour that they cannot be mistaken. The ground is pure white, distinctly but closely freckled with dark brownish purple; in some there are spots of a light purple underlying the others, but not easily seen except on close inspection. Some have the spots very thickly distributed, giving quite a darker tone to them, and being still more closely accumulated at the larger end. Others are thinly marked, while I have seen one in which the spots were arranged in a somewhat indistinct zone. The bird itself is elegant in shape, and the colours of the plumage are very pleasing, while its singular tremulous call cannot remain unnoticed.

The Willow Wren (*S. trochilus*) is much more abundant than the wood wren, and frequents the alder and willow trees growing on the banks of the streams, and the hedgerows of the adjacent meadows. The brook below the town is bordered here and there by rows of dwarf willows and hedges of the same, and here the willow warbler is numerous. Often while fly-fishing on a summer's evening I have been interested in watching their lively and active habits as they climbed and hopped about from twig to twig; now searching the river bank, now darting out to share with the trout below, the gnats that hovered over the water, and again regaining an overhanging bough they would thread their way quickly into the willow above, prying under every leaf and into each crevice for aphides and caterpillars. They do not always agree kindly with their fellows, but will chase each other in a quarrelsome manner, and occasionally direct their puny attacks on other birds.

The Chiffchaff (*S. rufa*) haunts the same spots, both species being often seen together; indeed, they are so much alike that an ordinary observer would not detect the difference, and I believe they are frequently confounded; but the legs of the chiffchaff are darker than those of *S. trochilus*, and the yellow mark over the eye is less distinct. Their nests, too, are placed in similar positions, and are formed of like materials, but the eggs are very distinct, those of the willow warbler having the ground of a pinkish white, closely freckled with light rusty brown, while the chiffchaff's have the ground a pure white, and are very sparingly speckled with

G

dark brown spots, more numerous towards the larger end.

The majority of our woods are oak and ash, chiefly the former; but here and there a plantation of Scotch fir and spruce is tenanted by that tiny monarch, the Goldcrest (*Regulus cristatus*). Though often overlooked because of its diminutive size, its sprightly habits make it worthy of attention, and with a little caution it may be safely approached without exciting its alarm. It is constantly in motion, like the titmice, and assumes every possible position.

It is singular how so small a bird survives the rigours of our cold season; but it really is very hardy, and in the depth of winter may be seen busy as ever, searching for its daily food, as if constant motion was absolutely necessary to maintain its bodily warmth; indeed, I do not remember ever seeing one indulging in the luxury of rest in the daytime.

It is my impression that they are more numerous with us in the winter than the summer, as if we received a partial migration from the north at that time; but I may be mistaken.

Its compact nest, formed chiefly of moss, is neatly suspended under a bough of a spruce or fir, and requires a sharp eye to discover it.

The titmice are a very interesting family of birds, and, though little in size, they seem determined not to remain unnoticed amongst their neighbours. Their constant activity and grotesque attitudes make them very amusing, and though some of them do not escape the censure of the gardener, yet few of our feathered friends are, I believe, more truly beneficial to us. Insects form the staple of their food, and, from their incessant care in feeding their numerous progeny, their consumption of the various ravagers of our fruits and flowers must be considerable, more so, perhaps, than those who have not watched their exemplary attention to their young would be inclined to believe.

Of the seven British species five are constantly to be met with throughout the year, though some are more abundant than others.

The Great Titmouse (*Parus major*) delights in a woodland home, but in the winter it is a constant visitor to our gardens. In its search for insects it is undoubtedly very destructive to the buds of fruit trees, and I have often remarked the partiality which it evinced for the buds of a Siberian crab-tree in my own garden. I have seen them in the beginning of December in our woods in company with *P. ceruleus*, *P. ater*, and *P. caudatus*, busily searching the mossy trunks of the old oak trees, prying into every crevice of the rugged bark, or clinging to the branches and plucking off with a vigorous twitch the withered leaves that still clung closely to them.

A writer in the *Gardener's Chronicle* states that he has observed the great tit come down on the roof of his wooden shed over his beehives, and tap on it with his bill until a bee came out, when he pounced on it, and ate it; and this not once or twice.

In our gardens the Blue Titmouse (*P. ceruleus*) is the most constant visitor. Ever in motion, it seems the personification of mischief, a veritable ornithological mountebank, for in the course of five minutes it will go through all the postures and attitudes which it is possible for a bird to practise, and while so doing it seems to have no fear of determination of blood to the brain, for it is quite as often seen with its head downwards as otherwise. It is a carnivorous little fellow, and delights in a bit of carrion. My next door neighbour, being a sportsman, kept a number of dogs, and to feed these, the carcases of sheep that had died in the fields were often skinned and hung up on poles around the kennel. These I have sometimes seen covered by as many as forty or fifty of the blue titmice, pecking away with all the vigour of which they are capable, and that is not a little. Tallow scraps, too, which were used for feeding the dogs, were much relished, but the carrion had the preference.

Professor Buckman has recently noticed that the blue tit benefits foresters by destroying the flies which cause the oak galls, which in many parts of the country are threatening ruin to young oak plantations.

The blue tit is not afraid to enter houses, and I have very often found them in a detached room in my garden that was used as a schoolroom, taking advantage of the door being left open. They would generally fly to the window on any one entering the room, but did not exhibit much fear, and when I have caught them in my hand the little things would bite fiercely at my fingers and try to effect their liberation.

Mr. Hewitson mentions a pair of bluecaps having built their nest in a bottle, and the following is another instance more remarkable still, and is well authenticated:—

In 1779 a pair of these birds built their nest in a large stone bottle that had been left to drain in the lower branches of a plum tree in the garden of Calender, near Stockton-on-Tees, and safely hatched their young. Every following year the bottle was frequented for the same purpose and with a like result, the bottle having been allowed to remain by the occupiers of the farmhouse. In 1822, the old plum tree upon whose boughs the bottle had been placed having fallen into decay, the bottle was rested on the branches of an adjoining plum tree, and fastened by iron hooks. The change of position, however, did not cause the little creatures to desert their home. The *Cumberland Paquet* of May, 1844, recorded the continuance of the tenancy without intermission up to that date. From another reliable source I am able to add, that in 1851 the pair made their appearance as usual to take up their summer residence. It had always been the custom of the inmates of the farmhouse to draw the nest of the previous year out of the bottle, but this year they had neglected to do this, and the pair selected another place for their nest. However, in the following year the needful preparations were made, and the birds again built their nest in the old domicile, and safely reared their numerous progeny.

Whether at the present time the bottle is still occupied I am not able to say. It would be an exceedingly interesting fact if we could ascertain how many pairs had tenanted the bottle during these years, for we may reasonably conclude that it was not the same pair. Taking the number of the young of this species at ten, which is, I believe, a fair average, between seven and eight hundred individuals must have been the produce of this "inexhaustible bottle."

The chastely-coloured Cole Titmouse (*P. ater*) is far from uncommon, but it is exclusively a denizen of our woods and plantations, and I never saw it in other situations. It does not refuse to mingle with others of the family; but it is in little parties of its own species, probably the summer's brood, that I have chiefly met with it.

Though insects form a considerable portion of its food, yet it is more a ground feeder than any of the others, and in the winter season it busily searches amongst the withered leaves for seeds, especially under the beech trees, the nuts of this tree being a favourite food. It is also very partial to the seeds of the birch, on the long pendulous twigs of which it clings in almost every position, swinging about with each passing breeze. Its more terrestrial habits are also shown by the position in which it places its nest, a hole in a bank or under the roots of a tree being often chosen for that purpose. I have heard its monotonous note as early as the 24th of January.

In our neighbourhood the Marsh Titmouse (*P. palustris*) is hardly so abundant as the cole. It is not by any means confined to low or marshy situations, for we have few such around us, but I have met with it far from water. The notes uttered by the great tit are like the whetting of a saw, but this is far more correctly imitated by the marsh tit; indeed, I have often been surprised at the close similarity, and have been tempted to look round for the sawyer. Their cry resembles the words "Chika, chika, chika," repeated four or five times in succession, and ending with a shorter syllable, "chike." Its habits are much the same as those of the others of the tribe, perpetually in motion, seeking its food in the crevices of the bark of the trees which it frequents; but, as far as I have observed, it does not associate with the other species, but keeps together in small parties.

The long-tailed Tit (*P. caudatus*) is the last I have to notice, for, as far as I know, neither *P. cristatus* nor

P. biarmicus have been seen in our district. The "bottle tit," as it is most commonly called with us, is very plentiful in our woods and plantations, particularly in those where there is a growth of underwood, in which it delights to place its nest. I have found it especially abundant in a large wood called the Catwins, on the outskirts of Thoresby Park. I have never met with a nest at a greater height from the ground than about four or five feet, nor is it at all particular as to concealment. I have most commonly found it placed in the fork of a young hawthorn, and on two occasions, where the fork consisted of three stout twigs, they were all included in the body of the nest, the moss and wool being so closely and firmly fitted around them that it was utterly impossible to detach the nest from its supports without completely pulling it to pieces. One of these especially excited my admiration, for the three branches, springing upwards equidistant from each other, were equally included in the structure they supported, which was woven of green moss intermingled with wool, and decked outside with grey lichens, the whole presenting a beautifully symmetrical appearance. The opening was near the top, and, as is usual, the interior was almost filled with feathers.

It has often been a wonder to me how the parent bird manages amongst such a mass of down to find all her numerous and tiny young ones. The opening to the nest is so small that when the bird enters, the interior must be almost in perfect darkness, and the marvel is that some of the gaping mouths below are not left unsupplied with food. But in this, as in all of our Maker's works, the means are perfectly adapted to the end, and if in any case we are unable to comprehend how this end is attained, we cannot withhold our faith in that Divine wisdom which both plans, and carries the plan into effect.

The pretty Bohemian Chatterer (*Bombycilla garrula*) has several times visited us. The winter of 1850 was particularly marked by the appearance of several flocks, chiefly during the severe frost in January of that year; many were shot, and all of these had their craws filled with holly berries.

The confusions of nomenclature were never, I think, more strikingly shown than in the family at which we have now arrived, that of the wagtails. Two or three specific and vulgar names are often applied to the same species by various authors, until it is extremely puzzling to make out which is meant.

The pied, which was formerly considered to be the *Motacilla alba* of Linnæus, but was found by Mr. Gould to be distinct, was changed to *M. lotor* by Professor Rennie; it is now, however, thoroughly established as *M. Yarrelli (Gould)*.

The white, which is the true *M. alba* of Linnæus, is called *cinerea* by Latham, while Montagu says the name of *white* wagtail, is a name for the *winter* wagtail, which he then describes under the specific name of *M. boarula* (*Linn.*), which is known as the *grey* wagtail, called by Macgillivray the "grey and yellow wagtail," and by Bechstein *M. sulphurea*.

There are, in fact, three species, which are often confounded under the trivial name of the *yellow* wagtail. The first is the one just named as the grey wagtail (*M. boarula, Linn.*); the second is the grey-headed wagtail (*M. neglecta, Gould*) the *M. flava* of Linnæus, and *Budytes flava* of Macgillivray; and the third is the true yellow, or Ray's wagtail, which by Cuvier was removed into another genus, and called *Budytes*, but which in Orr's edition of Cuvier (1849) is stated to be the *M. neglecta* of Gould, although previously known as *M. flava* of Linnæus.

According to the most recent arrangement they therefore stand thus : Pied wagtail (*Motacilla Yarrelli, Gould*), White wagtail (*M. alba, Linn.*), Grey wagtail (*M. boarula, Linn.*), Grey-headed wagtail (*M. neglecta, Gould*), Yellow wagtail (*M. flava, Linn.*). All these five I am convinced are clearly distinct species, and I have met with all of them in my own neighbourhood; even the rarer ones more than once, and under favourable circumstances for recognition and identification.

The Pied Wagtail (*M. Yarrelli, Gould*) is by far the most abundant of the family. Some of them remain with us all the year, but it is in spring and summer that they are met with in the greatest numbers. Their elegant form and active habits are very pleasing, and all their motions are marked by an airy gracefulness. During the winter they frequent the neighbourhood of houses, and I have often seen them busily employed in searching the roads and gutters.

They are very hardy, being far less affected by cold than many other birds who are apparently better able to bear it. On Feb. 17, 1855, during a severe frost, with the thermometer standing at 22° at the time, I watched with much interest one of these birds bathing in a shallow and rapid part of the stream running through the village, a few yards below the mill. It walked in as far as it could, and then with great energy dipped in its head and threw the water over its back several times, and with such evident enjoyment that you might have imagined it to have been July instead of February. Having finished its ablutions it bounded along the margin of the stream, apparently feeding eagerly, as if the bath had sharpened its appetite.

The ground is its favourite place, but in the winter it often perches on the tops of houses and outbuildings, seldom remaining there long, but flying off with an elastic bound and a cheerful twitter. Occasionally I have seen them alight on the rails of a fence, and once watched one feeding a young cuckoo, which was evidently its foster-child, both being perched on the top rail of a fence by the side of the stream. I never but once saw them perch on trees.

In March they are in some measure gregarious, for, though rarely associating in more than twos and threes in the daytime, yet in the evening they assemble in flocks of forty or fifty in number on the gorse coverts on the forest. They arrive in pairs and small parties about an hour before dusk, and perch on the bushes, continually shifting their places and uttering rather clamorously a shrill "t-wee." Often have I stood concealed and watched their proceedings, and as I listened to their busy twitter, I could fancy that they were each of them detailing their personal adventures during the day. As darkness drew on the gossip gradually ceased, and one by one they dropped down amongst the furze bushes, where they roosted for the night.

A bank at the bottom of a hedge I have found the most usual place for the nest, and frequently at a great distance from water. The foundation is generally formed of grass and roots, and lined with hair and wool, and rarely with feathers; cow's hair, which is doubtless picked up on the pastures, is the most usual lining.

Sometimes I have seen it used in such abundance as to form a mass more than half an inch thick, slightly felted together. I have known of two very singular positions for the nest. One of them was underneath a rail on a colliery railway which was in constant use, the parent bird flying off on the approach of a train of coal waggons, and resuming her seat on her eggs when it had passed. The other was built in the bows of a ferry-boat, and though the boat was constantly passing backwards and forwards, the young were successfully reared.

The White Wagtail (*M. alba, Linn.*) is I believe not such a stranger to our island as has been supposed. I have met with it several times, most frequently in the autumn and winter months. During the former season they have appeared in some years abundantly, while in the winter I have never seen more than one or two at a time. They were very numerous in September, 1854, frequenting the roads and margin of the stream running through the village. In January, 1855, during a sharp frost, and when the ground was covered with snow, I saw on two occasions a single bird in the street opposite my own house; it was busy searching the gutters in company with two or three of the pied species, and the distinction between the two was strikingly seen, the bluish ashen grey of the back of the white, contrasting strongly with the dusky colour of that part in the pied bird.

The Grey Wagtail (*M. boarula*) generally visits us in the winter, though it is not common. I have met with it both in the neighbourhood of houses (my own court-yard, for instance) and on the shallow parts of the stream, and in severe frosts I have even seen it wade into water as deep as it could bottom. It generally leaves us in March or the beginning of April, in some instances partially attaining the black feathers of the chin.

I have had the pleasure of meeting with two specimens of the Grey-headed Wagtail (*M. neglecta, Gould*). One was in my own courtyard in the summer of 1855, where I saw it several times during the day. The absence of any yellow on the rump, the darker colour of the legs, and the conspicuous black mark under the chin enabled me readily to distinguish it. The other occurred a few days later, and when I saw it, was busily engaged in feeding by the side of the stream in the village where it is crossed by the bridge, below which the water spreads out over a gravelly bed, and is very shallow. Here it was running nimbly along the edge of the water, and sometimes into it, rapidly seizing small aquatic insects, and twice it flew up and settled on a large stone in the middle of the stream, which was so far below the surface that the little bird looked as if it was swimming; but it seemed to have no fear of being carried off its legs. I watched its sprightly movements for nearly half-an-hour with great interest, and all the more so from its being a rare species. It often approached within a few yards of the bridge on which I stood, but at length flew away. I have little doubt these were a pair, as from the duller tints of the one I have last mentioned, I conjectured it to be the female. I looked in vain during the summer for their reappearance.

But if the last two species are somewhat rare, the Yellow, or Ray's Wagtail (*M. flava, Linn.*), is by no means so, but is constantly to be found in the meadows during the summer. Though not nearly so long as the three first named species, yet to my mind it is the most elegant of them all, its movements being very light and graceful. Though haunting the margin of the streams, it does not appear to enter the water so freely as the others, but seeks its food on the grass, on which it is well fitted for running by its much longer hind claws. At the same time I have remarked its fondness for frequenting the beds of the water daisy, which in summer nearly fills the stream with its waving masses, and where the birds appear to find a rich feast of aquatic insects. The pied wagtail is also constantly to be seen on these fish-beds, as they are called.

I have once met with the yellow wagtail in the winter —viz., on February 8, 1848. It was a solitary bird in a meadow near my own garden, where it was feeding by the side of a small carrier which takes the overflow from the stream above.

Amongst our common summer birds is the Tree Pipit (*Anthus arboreus, Bech.*). It is a favourite bird of mine, and in my solitary wanderings in the woods, its brief and singular flight and sweet song have often afforded me much pleasure; its habits are rather shy, and I never saw more than a pair together. I have found it most frequent in the wooded parts of the forest, not amongst the plantations, but where the giant oaks are interspersed with the graceful birch. Its favourite perch is a withered limb of one of the old veterans, springing from which it soars upwards in the manner of the sky-lark for about twenty or thirty yards, describing a half spiral in its flight, when it descends diagonally on outstretched wings and tail to the branch which it left. It is during its downward flight that its song is uttered, and sometimes, though but rarely, from its perch. With us it is seldom, if ever, met with in the cultivated parts, and it is amongst the grass, and moss, and heath of the forest, overshadowed with broad fronds of the fern, that it delights to place its nest.

There is little difficulty in distinguishing the tree pipit from the meadow pipit, though they have been frequently and strangely confounded. The former is so much more graceful and elongated in form, the ground colour of the neck and breast is more fawn than the latter, and the markings are more distinct, that it may be recognised at a glance by one who knows both species. On a closer inspection, the short hind claw of the tree pipit is an unerring distinction, as well as the lighter colour of the legs. Their habitat, too, as far as my observation goes in our own district, is as distant as their names imply, nor have I ever met with the two together.

The eggs of the tree pipit vary more than those of any bird I know, hardly any two being alike either in colour or markings; every tint from dark bluish-purple to rich red may be met with. Half a dozen are now lying before me. The first has a pale purplish-grey ground with very dark bluish-purple marks and blotches sparingly distributed, except at the larger end, where they are thickly accumulated; the second has a still paler ground, with blotches of very light purple, as though washed on; over this are spots of rich red, interspersed with smaller spots and lines of the same colour, but much darker, and crowded like the first at the larger end; the third has a pale reddish ground, pencilled over irregularly with a darker shade of the same, the larger end being also darker; the fourth is similar in markings, with the addition of distinct dark spots, the edges of which are somewhat shaded, but both ground and markings have a more purple tone than the preceding one; in the fifth the ground is pale reddish, minutely speckled all over with a darker shade of the same colour, but allowing the ground to be seen. The sixth has a still redder ground, but is so minutely freckled as to appear at a little distance of a uniform red. The six I have thus described I selected for my cabinet out of a large number I had collected, but they all varied so much that I had great difficulty in choosing such as I wished to retain as specimens. The variation, too, extends to the shape, some being rather short, with the small end very pointed, while others are more elongated, and some again almost oval.

The eggs of the Meadow Pipit (*Anthus pratensis*) have a brownish-white ground uniformly marked all over with minute specks of hair brown; the only variation is that the general hue of some is darker, from the specks being more thickly distributed. Montagu says that some are tinged with red, but I never met with such.

The Titlark, as it is commonly called with us, is a constant resident, but it is my impression that our numbers in summer are much greater than in winter. It is partial to cultivation, and its nest I have usually found in the meadows, placed on the ground, sometimes at the foot of a tussock of grass or a tuft of weeds.

The Skylark (*Alauda arvensis*) is as abundant with us as it is everywhere else. I do not think one of our native birds has so cheerful and inspiriting a song; it seems prompted by the very exuberance of joy and gladness, as if it could not be contained or controlled. What wonder, then, that both poetry and music should have chosen it for its theme; it would, indeed, be an almost endless task to enumerate the poets who have written on this delightful songster.

Its habit of singing in mid-air adds indescribably to the charm of its melody ; now its notes die away in soft cadences, now they come swelling in ringing glee. Mounting upwards, it leads our thoughts away from earth, and while we watch the tiny speck in the blue sky until it fails our sight, the notes of joy still fall on our delighted ear, prompting our hearts to rise in unison of praise to Him who made us both.

> "Higher still and higher,
> From the earth thou springest
> Like a cloud of fire;
> The blue deep thou wingest,
> And singing still dost soar,
> And ever soaring singest."

The skylark is one of our earliest songsters, even cheering the winter with its melody. On the 22nd of January, 1854, while the sun was shining brightly, I heard two singing as gaily as in summer, and another on the 10th of February the same year. It will sing also when everything is shrouded in darkness, as if the daylight was not long enough for its lays of love. On the 12th of April, 1853, very early in the morning, when it was so dark that I could not see distinctly many yards before me, and in the space of half a mile, I counted six or seven larks soaring at a great height, as I judged by their song, for of course I could not see them. About half an hour after this, the first faint tinge of light appeared in the eastern sky, and as it increased until first one object and then another came into view, bird after bird rose from the dewy grass with sprightly song, until the very air was vocal.

Our winter flocks vary greatly in numbers; in some years they are much more abundant than in others. In January, 1850, this was particularly the case; the frost in that year was very severe, and during its continuance the larks frequented the turnip-fields and fed on the tender shoots of the tops, as well as on those parts of the roots themselves where the sheep had bitten. Of this I satisfied myself by frequent observation. They never, however, assemble in such immense numbers on the downs of the southern counties.

I have not had the pleasure of seeing the skylark remove its eggs, as it is reported to do, but on two occasions I have known a nest laid bare by the mowers in my field, and on visiting each a few hours afterwards, the eggs were gone. No one had been in the field, and though in the case of one which was exposed, some prying crow might have abstracted the eggs, yet as there were no fragments of shells around, this did not appear to have been the case, while the other was almost concealed from view by the swathe of clover which partly projected over it. In both cases the eggs disappeared, and I have little doubt were removed by the bird itself. Any one who will take the trouble to place an egg in the foot of a skylark will find how easily it is clasped by the toes and their long claws, and what facilities these offer for its safe removal.

I have only met with one other member of this family —viz., the Woodlark (*A. arborea*), and it is by no means common. I have seen it often enough to be well acquainted with it, and have watched its flight, so different from any of its congeners; but it is sufficiently rare to be very interesting when it does occur. I once found its eggs, which were of a whitish ground colour, rather

H

sparingly speckled with brownish grey, except at the larger end, where they accumulated, and on two of them formed a very distinct and well-defined zone.

Of the family of the buntings I can enumerate five—viz., the snow, the common, the black-headed, the yellow, and the cirl. The Snow Bunting (*Emberiza nivalis*) is only a straggler with us; I have occasionally met with them during the winter, mingled with skylarks in the fields on the edge of the forest at Edwinstowe. Some individuals have been killed wearing the adult white livery, while others were in that immature plumage which has led them to be classed as a separate species, under the name of the tawny bunting.

The snow bunting migrates regularly during the winter, appearing in large flocks on the shores of the Humber; but I have not seen it in numbers to the southward of this boundary, those occurring in our forest district being but stragglers from the main body. The proportion of adult males in these flocks is but small, the majority being either females, or the young of the first year in the tawny livery. At times the numbers to be met with on the Humber banks are very large; they feed on the seeds of the dog-grass, the crops of those I have killed being literally crammed with them. They run along very actively, moving each foot alternately, and in the situations I have mentioned are very fearless, allowing you to approach within a few yards. If the weather continues severe, their visit is of some continuance; but no sooner is any indication felt of a change of temperature than they depart at once for their northern homes.

The Common Bunting (*E. miliaria*) is very plentiful on our arable lands, where its nest is placed on the ground, and generally with very little attempt at concealment. This species seems to be subject to variation of plumage—chiefly a large admixture of white—those I have seen having a dappled appearance; but I met with one in December, 1859, at Clipstone, which was entirely white, with the exception of two or three slight markings of brown on the back.

The flight of the common bunting lacks buoyancy, and consists of a series of undulations caused by the momentary closing of the wings, alternating with a few somewhat laboured flappings. It is by far the largest of the family, and is not by any means to be despised when well cooked.

In suitable spots the Black-headed Bunting, or reed sparrow (*E. schœniclus*), is frequent; the change from the dusky hue to the deep velvety black on the head of the male, is one of the earliest signs of the approach of spring, and in this, his nuptial dress, the male is really a handsome bird. I have always found its nest on the ground—most frequently near the bank of the stream, sometimes at the foot of a bush, at others amongst reeds and coarse high grass—but I never saw any attempt at suspension.

The eggs do not offer much variety; the ground-colour is generally a pale dirty brown, with a bluish or purplish tinge, and marked with distinct spots and curved lines of blackish or purple-brown, chiefly at the larger end; their shape is very similar to those of the yellowhammer, but the smaller end is rather more taper.

The vocal powers of the black-headed bunting do not attain to the dignity of song; two or three short notes, followed by one rather prolonged, in the manner of the yellowhammer, but harsher, being all its accomplishments.

But the Yellow Bunting (*E. citrinella*) is by far the most common species; it throngs every hedge and bush, and you cannot go many yards in the cultivated districts without seeing several. It delights to roll itself in the dust on the roads in summer time, and with such vigour that it raises quite a cloud. Its flight is a very broken one, a mere series of flittings, seldom continued for any distance. It is a handsome bird, though subject to variation in the markings, some having the yellow of the neck and breast much brighter and more unmarked than the others. I have seen one which had the whole of the head and neck a bright, clear yellow, entirely devoid of the usual olive-brown markings—most likely a sign of age. Its call, for it cannot be entitled a song, is very monotonous, and is well described by Bechstein by the syllables " tee, tee," repeated rapidly six or seven times, and ending with the more prolonged note "tchee."

It usually places its nest on the ground, the bank of a hedgerow being a favourite situation, and once or twice I have seen it placed on the thick lower branches of the hedge itself. The eggs sometimes differ remarkably in size, some nests containing one or two very much smaller than the rest. I have one I took out of a nest where the others were the ordinary size, which is only about half their dimensions.

The nest is composed externally of grass and fine roots, but internally it is a thick mass of hair, chiefly cowhair, and in form is very shallow. I have taken one which was the smallest possible remove from being quite flat, but the lining of hair in it was nearly an inch in thickness.

The Cirl Bunting (*E. cirlus*). This species, like *nivalis*, is not common, but they are occasionally taken on the forest fields at Edwinstowe during the winter, appearing with us as mere stragglers.

Amongst our native birds hardly one, I think, equals the Chaffinch (*Fringilla cœlebs*) in the exquisite construction and finish of its nest; and not one spends so long a time in its formation; I have known three weeks consumed in this process. It might well be thus when the elaborate style of the workmanship is considered, for indeed it is a very model of neatness; no straggling straws or other materials disfigure the symmetrical outline, but both the interior and exterior are perfectly compact and smooth. I have sometimes been led to believe that, in addition to the weaving and felting, by which the wool and moss and other materials are wrought together, the chaffinch uses its saliva for the purpose of increasing the firmness of its work. I have seen some of their nests which certainly appeared on removal, to have been attached to the branches of trees by other means than the mere weaving of the materials around them.

I was first impressed with this idea by finding a nest on the top of a post in my own garden. The post formed part of an open fence, on either side of which currant trees were placed; it was of split oak, and the top having been sawn off the surface was perfectly smooth, and nearly, though not quite, level. On this platform of six inches by four, without a splinter or projection of any kind to afford an attachment, I found in the middle of May a chaffinch's nest. It was such an exquisitely wrought specimen that I was tempted to remove it for my cabinet, and was astonished at the tenacity with which it adhered to the post.

The body of the nest was formed of wool, and lined with reddish cow's hair and two or three feathers; on the outside the wool was incorporated with green moss, and studded all over with green and white lichens similar to those on the rails of the fence; these lichens were more numerous towards the base, forming a sort of lip, and adhering to the surface of the post, on which there were no lichens growing naturally. Finding it cling so closely I used great caution in its removal, and am quite convinced that its adhesion was effected by means of some glutinous substance, most probably, as I have said, the saliva of the bird itself.

The male chaffinch in his brightest breeding plumage is an elegant little bird; some in this respect far outshine their fellows, but these perhaps are of more mature age. The clear bluish grey of the head and nape, the pink breast, and chestnut brown back, harmonize well with, and are set off to advantage by, the black and white of the wings. The males have a very distinct crest, which is raised and depressed at will.

They are resident through the whole year, but I have not noticed that marked separation of the sexes, as on the Continent and in the northern parts of our own island, to which it owes its specific name.

The local name given it is "spink," which is derived from its own well known note.

Though the chaffinch is chiefly a vegetable feeder, yet at some times of the year insects enter largely into its daily food. I have on several occasions seen it cap-

UNIV.
CALIF.

W.J.Sterland, del.

Vincent Brooks, Day & Son, lith.

turing flies on the wing, springing up in the manner of the flycatchers, and again returning to its perch. I once watched a female clinging to the wall of my house, and apparently employed in picking out insects from the joints of the bricks; she was thus engaged a considerable time, shifting her position easily, and using her tail as a fulcrum, in the manner of the woodpeckers. They are very fond of the tender leaves of radishes when first peeping through the ground, and often cause much annoyance by their depredations, but they are good friends in other ways, freeing us from many insect foes.

The pretty Mountain Finch (*F. montifringilla*) is a constant winter visitor, chiefly frequenting the beech woods, where it feeds upon the mast. Sometimes a straggler may be seen associated with linnets, but they generally visit us in small flocks, consisting only of their own species. The abundance of beech trees in Thoresby and Rufford Parks affords plentiful supply of mast; there the mountain finch is found in varying numbers. They do not exhibit much shyness, but permit themselves to be approached within a few yards while they are feeding; I have even seen them come close to the house without showing signs of alarm. They generally leave us about the middle of March for their northern breeding grounds.

I was for some time unaware that the Tree Sparrow (*Passer montanus*, Ray), was an inhabitant of our district. I had often found nests with their eggs in hollow trees, but I had always considered that they were those of the house sparrow. Having, however, shot one of the owners of a nest which I found in a cavity of a pollard willow, I saw at once my mistake, and recognised it as the tree sparrow, the dull chestnut of the head and nape forming a clear distinction from its relative, the house sparrow. Further observation showed me that it was more abundant than I had supposed; indeed, so much so that I cannot call it very rare.

With us it exclusively inhabits the cultivated districts, the meadows and hop grounds being much resorted to. As far as I have noticed, its nest has generally been placed in the hollows of pollard willows, of which numbers grow along the banks of the stream; the old oaks in the forest offer innumerable cavities in their decayed arms and trunks, but I never saw the tree sparrow avail itself of them; nor indeed have I ever met with it in woods.

Its habits are more shy than those of the house sparrow, and though easily recognised as a sparrow, yet its general form has a more graceful outline, and it is rather less in size. With the robust form it also lacks the pert impudence of its congener; and even in winter I never saw it mingle with the flocks of the latter which throng our farm and stack yards at that season. Its ordinary call is similar to that of the house sparrow, but shriller in tone; and it sometimes utters a few consecutive notes which are meant for a song, but have not much music in them. The eggs have a dull whitish ground, rather finely speckled all over with greyish brown. They do not vary much either in size or markings, though now and then I have found an egg in which the usually close speckles were replaced by larger markings and spots, sparingly distributed.

No bird is so well known or so universally distributed throughout the British Isles as the House Sparrow (*Passer domesticus*). Town and country, smiling fields and barren moorlands, wherever there is a human habitation, however humble, all are the same to him, for he is always at home. Everywhere he is the same fearless, independent bird; but the town sparrow is a much more pert little fellow than his brother of the country, and of the former class the London bird is the beau ideal—he seems to have borrowed all the forwardness and impudence of the London gamin, and as for fear or timidity, it has no place in his disposition. But it is with country sparrows that we have now to do; and though they are rather more unsophisticated than those inhabiting our towns, they are still a fearless tribe, and very amusing with their consequential and impudent airs.

But notwithstanding all that can be alleged against them, they are eminently serviceable to man, and certainly do not deserve the indiscriminate attacks which are made upon them. I believe the benefits they confer in the destruction of caterpillars and other insects injurious to our various crops, outweigh tenfold their consumption of corn and seeds, and I have found them most valuable assistants in the garden in clearing my gooseberry and currant trees of caterpillars; one pair of sparrows, during the season of feeding their young ones, will kill in a week more than 3000 caterpillars. I am convinced that the sparrow suffers unjustly from the many accusations brought against him by those who have not closely watched him feeding from one year's end to the other, but have formed their judgment from seeing, perchance, a flock revelling on the corn where laid by the wind, or even on the gathered sheaves.

Such an opinion I met with a little while since in the *Essex Herald*, in which the writer, after stating that the average yearly progeny of a pair of sparrows amounts to fourteen, goes on to say: "It is surprising that farmers should be so little interested on this matter; surely they cannot be aware that the little feathered tribe claim a tithe of their land's produce. The daily consumption of small birds is computed to be, in weight, one-sixth of their own bodies, and allowing the average weight of sparrows to be one ounce avoirdupois" (which, by-the-bye, is too high), "the consumption of 100 would be 6083 oz., or nearly $3\frac{1}{2}$ cwt. for the year. Supposing, further, that every hundred acres of land contained 1000 sparrows, their yearly consumption would be, according to the preceding theory, 60,830 oz., or nearly $34\frac{1}{2}$ cwt."

Now, undoubtedly, the consumption of this quantity of wheat or other corn would indeed be a serious matter, and if corn was the exclusive food of the sparrow, then something might be said in favour of his destruction. But we must not condemn him without hearing his own witnesses as well as those of his enemies. In the *Zoologist*, page 2349, Mr. Hawley of Doncaster writes that he has repeatedly watched sparrows feeding their young, and has found that on the average they bring food to the nest once in ten minutes for six hours out of the twenty-four, each time bringing from two to six caterpillars. He goes on to say, "Now suppose the 'three thousand five hundred sparrows'" (alluding to an association which had destroyed that number in a year), "were to have been alive the next spring, each pair to have built a nest, and reared successive broods of young during three months, we have, at the rate of 252,000 per day, the enormous multitude of 21,168,000 larvæ prevented from destroying the products of the land, and from increasing their numbers from fifty to five hundredfold!"

But we will leave estimates and suppositions for facts. It is well known that in France, where game is not preserved, a large class of "sportsmen" content themselves with shooting anything that comes in their way, and do not think it *infra dig.* to bag sparrows, linnets, and the like. The consequence is that small birds of all kinds have been so extensively destroyed that serious injury has resulted to the crops by the increase of insects, and numerous petitions have been presented to the Government praying that, on this ground alone, a law may be passed to prohibit the practice of destroying small birds. In one of the eastern departments the loss sustained in 1861 by the ravages of wireworm alone was computed at 4,000,000fr. or 160,000*l.*; and this enormous sacrifice of property was almost entirely caused by the ruthless destruction of small birds.

The attacks upon sparrows and other small insectivorous birds, however, still went on, with a consequent increase of insect pests; the agriculturists became alarmed at the result, and in June, 1864, presented four petitions to the French Senate praying for redress, and asserting that agriculture would "be seriously menaced if the destruction continued of their sole auxiliaries in arresting the propagation of insects, the scourge of all cultivation." In accordance with their prayer, a commission was appointed, presided over by M. Bonjean, which proceeded to collect evidence. The result as regards our friend the sparrow was, "that he, and he alone, could carry on the war successfully against the cockchafers and the thousand winged insects infesting the low grounds," and that in Hungary and in the Pays de Bade, where the sparrow had been exterminated, insects had increased to such an extent, "that the very persons who had offered rewards for his destruction, were the first to labour for his return, thus going to a double expense!"

I know something by experience of the numerous insects that ravage the gardens in Australia, and can understand the eager efforts made to introduce the English sparrow into that country. Our little friends took kindly to the climate and rapidly increased, and the following extract of a letter from a gardener near Melbourne, quoted by Mr. E. Wilson, shows clearly the value of their services :—

"A few weeks ago a portion of our grounds was literally swarming with caterpillars, and I dreaded the havoc that must ensue to our choice and valuable collection of young trees; fortunately, before any injury was done the sparrows came to our aid, not in scores, but in hundreds, and so completely destroyed the invaders that in less than ten days very few of them were to be seen; and at the present moment the sparrows may be seen all day long following up the trail of the caterpillars, and ravenously destroying the last remnants of the army that may have before escaped their vigilance.—Dec. 11, 1866."

It was in 1862 that sparrows were introduced into Australia, and so rapid had been their increase, that in 1868 the colonists were complaining that the fruit in their gardens had been largely destroyed, and alleging that the sparrows were the chief depredators. In April, 1868, the Secretary of the Victoria Acclimatization Society was directed by the council to write to Mr. Wilson to ask him to "assist them in procuring evidence as to the utility of the sparrow to the garden and the farm." Now to me it is simply marvellous, that with the actual evidence before their eyes of the value of the sparrow in destroying insects, they should make such an inquiry; as an old colonist, I feel ashamed of such an appeal *ad misericordiam*, because the sparrows had eaten a few of their grapes and cherries. Well, Mr. Wilson proceeded to make some inquiries, and amongst other information obtained this remarkable fact, that a gentleman had picked up below the nest of one pair of sparrows 1400 wing cases of the cockchafer! Now this insect, especially in its prolonged larval condition, is one of the most destructive enemies of the agriculturist, and in consequence of the practice I have adverted to, has increased to such a frightful extent in France, that Mr. Wilson says the damage they have done to the crops has been estimated in some years as high as forty millions sterling! That this is not an undue estimate will be seen from the following extract of a correspondent of *The Field*, dating from Havre, May 6, 1868 :—

"Gardening is here carried on under very great difficulties. Every Frenchman who has a chance is a 'chasseur indomptable,' and consequently great is the destruction of every kind of small bird, so that the insects enjoy a perfect jubilee. The air has been black with cockchafers during the last ten days. So great is the damage done by them that a penny per pound weight is paid for them, and numbers of men and boys are engaged hunting them. On Saturday, a cart drawn by two horses, threw its load of over 3000 kilogrammes of dead cockchafers into the sea."

Such is the result of the undue interference with nature's laws. No one asserts that the presence of the sparrow is an unmixed good, but the balance is so largely in his favour, that he ought to be welcome to take a little fruit or corn as wages which he has fairly earned.

In the face of facts like these, who will be inclined to hold up the sparrow and our other tiny feathered friends as hostile to the farmer, and what in this enlightened nineteenth century are we to think of the *intelligence* which could perpetrate such acts as the following letter records, and which I copy from the *Times* of December 12, 1862?—

"SPARROW MURDER.—I think the following exploit of the 'wise men' of Crawley ought to be shown to the world in your widespread journal; it speaks for itself, and requires no comment on my part. It is taken from a country paper of this week:—'CRAWLEY SPARROW CLUB.—The annual dinner took place at the George Inn, on Wednesday last. The first prize was awarded to Mr. J. Redford, Worth, having destroyed within the year 1467. Mr. Heaysman took the second, with 1448 destroyed; Mr. Stone third, with 982 affixed. Total destroyed, 11,944; old birds, 8663; young ditto, 722; eggs, 2559.—Yours obediently, A REAL FRIEND TO THE FARMER.—December 10, 1862."

I do not know where Crawley is, but I feel ashamed of the profound ignorance and inhumanity of its inhabitants, and especially of the three individuals who carried off the prizes in their sparrow club.

The nest of the sparrow is a loose, careless structure, and it is amazing to see in some cases the quantity of materials of which it is composed without any apparent necessity for such an accumulation. The mouth of a cast iron pipe, about six inches in diameter, proceeding from a stove in a laundry attached to my father's house, was, singularly enough, chosen every year, and sometimes twice in the season, as the site for a nest. The stove was only used every fortnight, and in this time the nest was built and some eggs always laid, but I never knew the parents bring up a brood, for the smoking of the stove always led to the obstruction being discovered and removed; and sometimes I have found the eggs quite baked with the heat.

The space underneath the tiles of my own house was generally occupied by a pair or two of sparrows, and hearing one day a very noisy commotion on the roof, and seeing numerous birds flying to and fro in apparent trepidation, as if some calamity had befallen them, I was convinced something was the matter, and procuring a ladder I mounted to the spot, and at once discovered the reason for all the outcry. One of the owners of a nest underneath the tiles—the female—in passing through the small aperture leading to her domicile, and which at the lower end tapered quickly, had evidently slipped, and her neck had become so securely wedged between the tiles that escape was impossible. Her dying struggles attracted her neighbours, who with great goodwill had done their best to extricate her from her unfortunate position; their zeal, however, was greater than their discretion, for they had pulled and tugged so earnestly that, when I arrived on the scene, hardly a feather was left on the body, which of course was lifeless.

I remember another similar instance, and on the same roof too, where a young one in leaving the nest had got its leg entangled in a loop of a piece of worsted which was amongst the materials composing the nest. It vainly tried to free itself, and, as in the former instance, a great crowd assembled to assist their unfortunate companion; but their efforts did more harm than good, for, as it hung halfway down the tile suspended by the thread, they had tried to release it by pulling it, and with the same result as in the other instance, for by the time I reached it, it was half stripped of its feathers, and its little life was almost gone.

In both these cases I feel convinced that the efforts which were made by the companions of the luckless sparrows were prompted by a feeling of compassion and a real desire to alleviate their misfortunes; their anxious hurrying to and fro, and the distress expressed in their cries, clearly indicated this. I have seen similar feelings of alarm and sympathy shown by domestic poultry, when on one occasion a cock was flying to the top of a fence in my own yard, but missed his aim, and fluttering down, his head slipped between two of the palings; the hens hurried to help him, but of course unavailingly, and he would soon have been strangled if I had not gone to the rescue.

I remember an instance, however, in which the circumstances were similar, but I am not quite so sure of the nature of the feelings which prompted them. In what is called the Dark Wood, in Thoresby Park, there are several old oaks growing on a high bank, from which, on the lower side, the earth has fallen away, and exposed the interlacing roots of the trees. This spot is much resorted to by the fallow deer, who, when the velvet is ready to fall from their newly-grown antlers, delight to hasten the process by rubbing them on these roots. On one occasion a fine buck with a full head was thus engaged when his horns became locked in such a manner as to be inextricable. All his struggles were in vain, for when found he was still fast, but quite dead, having been gored in numerous places by the antlers of his companions. Whether they had done this in hostility to one who may have rendered himself obnoxious—for there is a great spirit of rivalry amongst the bucks—or whether the wounds had been inflicted in kind but vain endeavours to effect his freedom I know not, but I am inclined to think the latter may have been the case, and that the friendly spirit had been manifested with more zeal and energy than judgment. Had it been in the autumn, during the rutting season, when fights are constantly occurring between rivals, it would have only been natural to refer it to the former cause.

I have often been amused to see a sparrow take possession of the nest of a house martin (*Hirundo urbica*). The eaves of a house near my own were always selected by the martins year by year for their erection, and rarely has a season passed without one of these aggressions occurring, which I have watched from my windows with much interest.

It always appeared to me that this forcible taking possession of their neighbour's house by the sparrows, was never done with the intention of making it their own residence, but from sheer mischief, and a desire to tease and tantalize the poor martins. These invasions always took place when the nest was empty, either before any eggs had been laid, or after the young had gained sufficient strength to take wing. I have watched the sparrow sitting quietly on the tiles above the nest, as if he was the most innocent creature possible, intent only upon his own affairs, and had not the slightest thought of intruding upon his neighbours; but the moment he became assured that the nest was unoccupied, he fluttered down, and popping in, turned himself quickly round, and sat with his head peeping out of the opening. Great, of course, was the consternation and distress of the martins on discovering the intruder, but though I have seen the incident dozens of times, I never saw any attempt to attack or eject the sparrow, nor, as it is asserted has been done, to stop up the hole with clay, and thus to inclose him, as the erring nuns were of old,

"Alive, within the tomb."

I have always been inclined to disbelieve this story, for I thought that the sparrow was too bold a bird to sit quietly and allow itself to be thus immured, when a few strokes of its strong beak would speedily demolish its prison walls; but Macgillivray adduces three such well-authenticated instances of a similar occurrence, that I am compelled to abandon my doubts in the face of so eminent an authority. One of these instances he thus relates:—

"A few years ago, in the window of a second story of a house in Linlithgow, inhabited by Mr. James Brown, buckle-maker, a pair of martins built a nest, which was taken possession of by a female sparrow. In attempting to dislodge this bold intruder, a dozen of their companions came to their assistance, but after many severe struggles they were unable to effect their object. For her rash conduct, however, they were determined to make her suffer. They agreed to entomb her alive by closing up the entrance with the mortar which they use in building their nests, and in this they succeeded. Mr. James Douglas, slater, with whom I have been a long time acquainted, and upon whose veracity I can depend, assured me that he was a spectator of the occurrence, and that he in the presence of several individuals, some of whom he named, took the dead bird out of the nest. The truth of it is further confirmed by Mr. John Ray, nailer, in Linlithgow, who told me he was also present when it happened."

What I have said about its boldness is well proved by the following incident, which occurred at the vicarage of Beeston, near Nottingham, in August, 1859. Numerous flocks of sparrows had frequented the grounds, and the cat belonging to the house had been watching their arrival, and seized every opportunity of pouncing upon them. She was at the foot of a tree one day looking up at the sparrows, and doubtless on murderous deeds intent, which they seemed to divine, for in a few minutes they descended *en masse*. As the birds came within reach the cat made a spring at them; but the tables were now turned, for so fierce and pertinacious was their attack, so closely did they follow up their enemy, hemming her in on all sides, that she was perfectly cowed, and compelled to seek safety by springing through a window, leaving the victory to her brave little assailants.

I knew a few years since an instance of the power of imitation which the sparrow possesses. A young one was brought up by a person at Newark from the nest, its place being always in a cage by the side of a skylark. Here it learnt the song of the lark, and would repeat it so accurately that if you did not see the bird it was impossible for a time to tell whether it was the lark or the sparrow that was singing. Often have I heard and admired its surprising imitation, when suddenly it would cease its song and utter the usual harsh chirrup of its race. Sometimes its sweet song would be frequently interrupted by this natural note, while at others it would sing for a long time without giving vent to it.

Variations of the sparrow's plumage are not uncommon, being chiefly interminglings of white. In December, 1859, one was shot at Ollerton which had the whole of the plumage white, the head and back merely having a slight tinge of brown, giving the white on those parts a dirty appearance.

I have seen a singular place selected for the nest of the sparrow—viz., the ornamental iron brackets supporting the roof over the platforms of several of the stations on the Liverpool and Manchester Railway, and where they seemed quite unconcerned by the passage of the trains.

The Greenfinch (*Fringilla chloris*) is a common bird with us in the summer, chiefly frequenting the cultivated districts, but in winter it is less abundant, or at least apparently so, and I have seldom seen it congregating in large flocks as the linnet does.

The vocal powers of the green linnet, as it is locally called, are very limited; its ordinary note, uttered chiefly when perching on the topmost spray of a hedge, is rather a melancholy one, and Meyer very correctly represents it by the word "tway." It is a shy bird, and at once flies off on your approach, or betakes itself to the tops of the trees, from whence they soon descend when the danger is past. It builds a neat nest, which is generally well concealed in a bush or hedge.

I have had the pleasure of meeting with the Hawfinch (*F. coccothraustes*) several times, but chiefly in the winter. The last occasion was in the winter of 1859-60, when a small party of four made their appearance in the shrubberies of Rufford Abbey. They arrived at the beginning of November, and remained for several months. Their habits were very shy, and they confined themselves to the clumps of evergreens, principally holly and box, where they appeared most frequently employed on the ground underneath the shrubs. One of them, a female, was seen on the 19th of March, when they appear to have left. During the same season another pair were seen at Cuckney, but both were unfortunately shot.

A hawfinch in immature plumage was caught in Thoresby Park in July, 1864; it had one of its wings hurt, which prevented it from flying, and consequently permitted a workman to take it up in his hand. He carried it to his workshop in the woodyard, and there offered it some green peas, which, to his surprise, it ate greedily, taking them in the most fearless manner from his fingers. In August following, when I saw it, its wing had healed, and it took well to confinement, but was very shy when strangers approached, fluttering to the further side of its cage, though it manifested no alarm at its captor, with whom it was quite familiar, and would take food from his hand.

The Goldfinch (*F. carduelis*) is at once one of the most beautiful, as it is one of our commonest, song birds. With us it is especially abundant on the forest, where, on the open parts, numerous plants of the common thistle grow, either singly or in patches; here I have often watched small parties of these pretty birds clinging to the prickly heads and rifling them of their downy seeds, incessantly uttering all the time their musical call notes, as if they could not contain their enjoyment. In our gardens it feeds on the silky seeds of the groundsel: it delights, too, to build its nest in such places; and a small clump of nut and plum trees in my own garden, overshadowing an arbour from which a honeysuckle clambered upwards and entwined them together, has often been selected by a pair of goldfinches, their nest being always placed on the bough of a plum tree.

The nest of this species is an elegant structure, very similar to that of the chaffinch, though if possible more elaborate in its compact felting of wool and hair, but I have rarely found any twigs used in its construction; the exterior varies with the situation and the materials to be obtained, sometimes being ornamented with moss, at others with lichens, the last being most frequent. The country people call the goldfinch the "proud tailor," and truly, in the construction of its nest, it may well be proud of its exquisite workmanship.

The natural song of the goldfinch, though sweet, does not possess much variety, but it is a good imitator, bears confinement cheerily, and is therefore much prized as a cage bird. If the young ones are taken before they can fly, the parents will feed them readily if put in a cage to which they can have access. The adult birds are very easily caught in a trap-cage, and soon become reconciled to their prison.

The Siskin (*F. spinus*) is a rare bird with us, at least as regards its visits; as a species it is abundant, and when it does make its appearance it is always in large flocks, but no one locality appears to have the preference, nor do its migrations appear to be guided by any fixed laws, but are fitful and uncertain.

In our district I have only met with it three times— twice in the winter of 1848, and again in February, 1854. It was in the first week in January in the former year that I saw a large flock of sixty or seventy, who were busily occupied in extracting the seeds from the berries of a group of alders in Rufford Park. I was riding near the trees when my attention was attracted by the birds rising from them. The action was a most singular one, for so simultaneous was the flight of the flock, and so exactly alike was the movement of each individual composing it, that it was just as if all were regulated by one will instead of many. In this compact phalanx they wheeled about for a few turns, uttering at the same time a shrill twitter, and again alighted on the trees and commenced feeding, suffering me to approach within half a dozen yards of them, and at this distance I watched them carefully with extreme pleasure.

While engaged in picking the seeds from the alder berries, they clung in every imaginable or unimaginable position, exactly like the blue titmice, of which they strongly reminded me, and, like that species, hanging with the back downwards as often as otherwise. After observing them for some time, I roused them from their employment with a stone, being curious to witness again their beautiful evolutions, which were performed exactly as before; they did not seem at all alarmed at my interference, but again descended *en masse*, and recommenced their occupation.

I spent some time in close observation of their habits, exhibited under such favourable circumstances, and was the more interested, from this being the first time I had seen them in a state of nature. I am sure that no one who had only seen the siskin in a cage would conceive the ease and grace of its movements, and its extreme activity when in freedom. I should like to have secured a few specimens, but had not the heart to fire a gun at the pretty little creatures, when a single discharge would have killed fifteen or twenty, so closely were they packed together.

I met with another, or perhaps the same flock, about a month later in the year, and not far from the place where I saw them before. In 1854 we were visited by a smaller party, who, like the others, were feeding on the seeds of the alder. These, with the addition of a single one which was shot in company with some linnets and brought to me, are all I have seen.

No bird is more common with us than the Linnet (*F. cannabina*). In summer it is scattered all over the heathy tracts of the forest and the cultivated fields adjoining, and even in the close vicinity of the village. In winter they assemble in large flocks in the stubbles, and I have seen them frequent corn stacks that were erected in the fields, clinging to the sides and picking out the corn. Notwithstanding this, they do good service to the farmer and gardener by feeding on the seeds of many troublesome weeds, such as the thistle, the dandelion, and others of the same winged character.

Though the heath and gorse bushes on the forest offer innumerable suitable sites for the nest of the linnet, it does not by any means confine itself to such places, but builds in the hedges also, where I have often found its nest. Bolton says the nest of the linnet is lined with "hair, wool, and the down of willows;" it may be so, but I never met with any other lining than wool and hair, with a feather or two.

In the next species, the Lesser Redpoll (*F. linaria*), the case is different. It is by no means a rare bird here; and I have generally found the nest placed in a low bush of alder or willow. One is now before me, and consists entirely of thin dry bents, woven together with wool, and thickly lined with the snowy down of the willow catkins; this forms a beautiful bed for the eggs, which are of a bluish green, speckled with orange-brown chiefly at the larger end, and are both smaller in size and deeper in the ground colour than any of the other species. Like the former species, they feed on the seeds of the thistle, dandelion, &c.

The Mountain Linnet or Twite (*F. montium, Gmel.*) is the last of the family I have met with, for I am not aware that the mealy redpoll has occurred with us. The twite is abundant on our heathy grounds, where it regularly breeds. The nest is generally placed in a tuft of heather, but I have taken one out of a furze-bush. It is composed of small roots and sprigs of heather, with here and there a dry bent, the whole being interwoven with moss, and lined with hair mingled with a few feathers. The eggs are numerous, sometimes as many as seven, the ground-colour being pale greenish white, with small dashes of light yellow-brown, and spots of purplish-brown, chiefly at the larger end.

When perching on the tall heather, or gorse, it keeps uttering a single note resembling its name, "twite," but when it flies off this is rapidly repeated in a twittering manner.

Our well-wooded district is a favourite one with the Bullfinch (*Pyrrhula vulgaris*), and its conspicuously coloured and portly form is consequently very common. Its provincial name with us is "Pick-a-bud;" and assuredly it is not undeserved, for it makes sad havoc in the gardens amongst the fruit trees in spring time. I am not inclined to agree with those who consider that the bullfinch and titmice destroy only those buds which contain a grub, for I have seen branches of gooseberry, cherry, and Siberian crab so entirely stripped of their buds year by year, that I cannot conceive such to be the case, or other branches of the same trees which were not so denuded would certainly show more traces of the ravages of the grub than they have done. I have watched them in a Siberian crabtree in my own garden, which stood about three yards from the window, and I feel convinced that they eat most of the buds they pick off; for the ground under this tree only showed one here and there which the birds had let fall.

The bullfinch is a permanent resident in the district. In spring it is only met with in pairs; in autumn and winter it associates in small parties of five or six in number, most probably the members of one brood. In winter it chiefly frequents the fields of stubble for seeds, and I have often met with it in hawthorn hedges feeding on the haws, to which it is very partial.

Mr. Morris, in his British Birds, conceives that the name bullfinch is a "corruption of budfinch, the word bud being pronounced in the vulgate of the north of England as if spelled 'bood;'" but surely this is a forced conjecture; is it not rather derived from the thick rounded form of its head and body, and its short neck? The word bull is used in many compound words to express largeness and roundness, as "bullfaced," having a large face; "bulltrout," a large kind of trout; "bull-rush," a large rush. This is the sense in which I have always been accustomed to consider the word, and have thought it very expressive.

CHAPTER IV.

PERCHING BIRDS—*continued.*

SEVERAL instances of the appearance of the Crossbill (*Loxia curvirostra*) have come under my notice. Many years since a large flock visited a number of Scotch firs and larches which grew around the house of a friend of mine in the village, although I cannot specify the exact year. Of course their rarity made them an object of attraction, and one of them was captured alive by my friend in a rather singular manner. The surface of a small pool of water in his stable yard happened to be covered with chaff and dust, which had blown upon it from a quantity deposited near; on this treacherous surface one of the flock, a male, descended, but the poor bird was speedily undeceived by sinking into the water, and so wetted his plumage that he was unable to rise, and became an easy capture. My friend put it in a cage, where it spent no time in unnecessary regrets, but cheerfully resigned itself to its confinement, being plentifully supplied with fir-cones, on the seeds of which it eagerly fed; here it remained for several months, when it was accidentally liberated by the servant.

On the 18th of February, 1856, a small flock of fifteen visited some elm and larch trees at Edwinstowe, where they were extracting the seeds from the few remaining cones of the latter, and biting off the buds of the former. Their motions greatly reminded me of those of the parrot tribe; they climbed with equal facility, holding on by beak or feet, and twisting themselves round the boughs in every possible position, all the while uttering a shrill twittering expressive of satisfaction. They remained in the immediate locality for a day or two, but alas! only to meet the fate of rare birds, for they were so intent on their occupation that the whole of the flock—twelve males and three females—suffered themselves to be shot.

Other birds were killed in March of the same year at Rufford, and in the following April some were also killed in a fir plantation called Ollerton Hills.

In the year 1849 Mr. H. Wells shot twenty-five on the firtrees surrounding the house of the late Lady Scarborough in the village of Edwinstowe.

It is rather singular, as noticed by Montagu, that the mandibles of the crossbill do not always cross on the same side. A pair of the flock mentioned above, which I obtained from the person who shot them, vary in this particular, the upper mandible of the male crossing to the right, and that of the female to the left.

In Macgillivray's account of this species he quotes Yarrell's description of a young one which was taken when only just able to fly, the mandibles of which were quite straight, the under just shutting into the upper, and then makes this curious remark: "It then appears that until the crossbill has used its beak in extracting the seeds from between the scales of the cones of pines and firs, so as by the peculiar action which it employs in so doing to bend the tip of the upper mandible to one side, the curious crossing and elongation of the tips of the mandibles characteristic of this genus are not observable, the bill being similar to that of a finch or sparrow, though stronger and more compressed."

Surely his meaning in the passage above quoted cannot be that if a young crossbill were taken before it left the nest and prevented from feeding on its favourite food, the crossing of its mandibles would never take place, but that they would remain straight, like those of a finch or a sparrow! Apparently it is, but to my own mind such an idea carries no weight, for I believe that the deflection of the mandibles would gradually be accomplished, even if the bird never tried their power on a fir-cone. The whole instrument, instead of being as Buffon declared it, a "useless deformity," is a most beautiful adaptation of means to an end, for when the points of the mandibles are brought together and inserted beneath the edge of a scale, the very powerful muscles by which they are moved across each other gives them a wedgelike action, which forces open the scales of the cone and liberates the seeds—a process which would be otherwise impracticable to it, and one which finches and sparrows never accomplish. The special development of the muscles of the cheeks in this and the allied species shows clearly the use for which they are manifestly designed, and is sufficient to dissipate the idea contained above.

Two other members of the genus Loxia have been taken in the district, both being rare stragglers.

The first, the Parrot Crossbill (*L. pityopsittacus*) is a native of the north of Europe, its stronghold being the pine forests of Norway and Sweden. It is but very seldom that any of this species visit our shores, but in the winter of 1849, a small party were seen in a clump of Scotch firs at Edwinstowe, and were all shot by Mr. H. Wells on the 4th March. Within a month from that date four of the American White-winged Crossbills (*L. leucopterus*) were shot in the same trees. The latter is a rarer visitor to this country than the former, and though a native of North America, being found in all the extensive forests of that continent, it yet appears to be sparsely distributed in Sweden and Norway, and it was doubtless from these countries that our visitors came, and not from America. Both species were busily engaged in feeding on the fir-cones.

The Starling (*Sturnus vulgaris*) frequents the old oaks in the forest by thousands and tens of thousands. Every tree during the summer has its several pairs of birds, who build their nests in the holes and decayed cavities in company with the jackdaws. In the autumn they collect together in immense flocks, and leave the district for the winter, resorting to the reed and osier beds on the Trent. Their return to us is very gradual, a few pairs being seen in some years as early as the middle of January, in others not until some weeks later.

Pairing has already taken place in those who reach us the earliest, and their peculiar guttural breeding-call I have heard at the beginning of February. Every week adds to the number until we receive our full complement, and the woods resound with their prolonged plaintive whistle, alternating with an oft-repeated gurgling note.

The starling is not only subject to local migrations, but I believe large flocks leave us for the continent in the autumn, and return in the spring; indeed, the fact that on one occasion seventeen dozen were picked up near the lighthouse on Flamborough Head, which had been killed, lamed, or stupefied by flying against the lantern of that brilliant light, seems to leave no doubt on the question, as they were evidently approaching our shores from the continent.

The great abundance of old decaying oaks in the forest leaves the starlings little to desire in the choice of a resting-place, and with us other sites are but seldom selected; but in 1853 I met with several pairs which had appropriated some deserted holes of the sand-martin at Robin Dam, near Rufford, a very unusual site. Pigeon cotes are also chosen, but in these cases the poor starlings become the victims of an ignorant prejudice, the common idea being that they suck the eggs of the pigeons. I fully believe them guiltless of such a habit, and that they are prompted to resort to such places only by a natural instinct to secure a comfortable domicile for their young. The stove-pipe, which I have mentioned as generally occupied by the sparrow, was once selected by a starling for its nest, but its eggs shared the same fate of being half baked.

No skill is exhibited by the starling in the construction of the nest, but it accumulates as great a mass of materials as the house sparrow, chiefly dry grass and straw, and rudely lined with a few feathers. I never met with it otherwise than in a hole or cavity of some kind.

Insects form the staple of the starling's food, and I think are always preferred when attainable. I have occasionally seen the birds seize insects on the wing, although it is not a common habit, and the first time it came under my notice it struck me as very unusual. A pair had a nest in the hollow of an old oak in the forest, which at the time contained young ones; one of the parent birds flew out of the hole, which it had just previously entered, and was rapidly departing for a fresh supply of food, when it suddenly deviated from its course and seized a large insect which was flying near, and then darted on one side and captured another. I was within four or five yards at the time, and had a distinct view of what was to me then a novel proceeding, but I have since observed it several times, and particularly so on the 20th of May, 1856, when I saw a number of them hawking for flies in the manner of the swallows. In this case it was no momentary impulse that prompted the habit, as in the first instance, but they were steadily making a business of it, and continued thus employed for some time.

They are staunch friends of the farmer, and consume an immense number of grubs and slugs, and in their search for these I have seen them literally blacken the pastures with their numbers.

The Raven (*Corvus corax*) is but a straggler in this part of the country. I have not known of more than two instances of its occurrence, a fact which I have been rather surprised at. Its predaceous character makes it many enemies, and I have never known it to nest with us. I have the evidence of old residents that it used to be comparatively frequent in the district, but it is very rare now.

Many years ago the landlord of the Black Bull Inn, at Mansfield, had a tame raven in his stable-yard. I always frequented this inn when I had occasion to go to Mansfield, and whenever I drove or rode into the yard, Tom, the raven, was sure to be about; and if the ostler was not in sight, he invariably called out with a hoarse but distinct voice, "Ostler, come and take the gentleman's horse!" bustling about all the time in a pompous, amusing manner, as if he had sole charge of the yard.

The Carrion Crow (*Corvus corone*) is a much more frequent species, but as its predaceous habits bring it unrelenting hostility, it is not abundant. Farmers dislike the "corby" for its attacks on their lambs as much as the keepers do for its ravages on their game, so its numbers are constantly thinned.

They are rarely seen together in greater numbers than a single pair, and these appear to remain constant to each other throughout the year. Yet though exhibiting much affection and faithfulness to each other, their omnivorous appetites do not tend to recommend them to our notice. Nothing comes amiss to them. The young of hares and rabbits, as well as the nestlings of any species of bird, are especially subject to their attacks, and they are particularly partial to the eggs of those birds that breed upon the ground, the partridge and the plover for instance. I have seen them in a place much resorted to by the latter bird, regularly hunting for their eggs, of which they are very fond, while their poor victims flew wildly about, uttering their "pease-weep" in a very disconsolate and distressed tone; the young of many birds are also greatly subject to their depredations.

The Hooded Crow (*Corvus cornix*) is a regular visitor during the winter months, from November to March inclusive. The earliest date of its arrival I have noted was the 21st of October, and the latest of its departure the 5th of April. Mr. Harley of Leicester, in a communication to Macgillivray (vol. i. p. 721) says: "An old coachman, who for at least twenty years drove the London and Leeds Express coach from Loughborough to Mansfield, across Sherwood Forest, used to say that he knew to a day when the grey crows would come upon the forest. That day he said was Guy Fawkes day, of notable memory, the 5th of November." My own observations would lead me to confirm this as the usual, though, as will be seen above, not the invariable time of their arrival.

The hooded crow generally frequents the uncultivated districts, chiefly the wooded parts of the forest and parks; and though you may always make sure of seeing it, it never occurs in flocks or large parties, but generally in pairs, or as solitary birds. This species appears to be only partially migratory, for though I never missed its presence during the winter, yet at the same time numbers are always to be seen on the seashores, its habits being essentially maritime. There it follows the ebb and flow of the tide with great constancy, feeding greedily on anything that may turn up. It is very shy and wary, rarely allowing you to come within gun-shot unless you do it cautiously, yet at times I have seen it exhibit much boldness and fearlessness. · I was riding on one occasion through the forest, where, perched on an old oak, sat a pair of these birds. As I came near, a sparrowhawk flew past within a few yards of the tree on which the hoodies were sitting, when one of them immediately took wing and attacked the hawk with such fierceness and pertinacity that he seemed fairly cowed. He made no show of resistance, but doubtless thinking discretion the better part of valour, left the field to his assailant, who, after following him for some distance, returned in triumph to his mate, who had remained on the tree a quiet spectator of the combat.

In February, 1855, I passed through a field where a dead horse was lying, on which two shepherd dogs were making their repast. A pair of hooded crows, attracted by the carrion, came sailing by, and alighted on the ground within a few yards of the carcase; both the dogs immediately ceased their feast and sprang at the birds with a growl, driving them off to a short distance. Nothing daunted, the birds returned, and again were driven off, and this was repeated many times, on each occasion the birds approaching nearer the horse until they settled down on the carcase within two feet or so of the dogs; but they were excessively wary, and if either of the dogs ceased eating, or raised his head, the crows sprang to the wing in a moment; and in this way they managed to secure a portion of the coveted food. It was during a severe frost, which perhaps had sharpened their appetites and rendered them more fearless than usual.

The food of the hooded crow during its inland sojourn seems to be chiefly carrion, worms, and grubs, and they likewise devour eggs. I once watched one flying near the island in Thoresby Lake, where a party of five herons sat preening their feathers, while a sixth was sailing overhead. On the approach of the crow the heron on the wing immediately gave chase, uttering shrill cries; hoodie, however, exhibited no boldness this time, but sneaked away without delay. The island is covered with tall trees, in which the herons build one or two nests each spring, but the eggs are usually destroyed by the carrion crow; possibly the hoodie was mistaken for one of his sable brethren, and hence the attack.

I met in a Scotch paper with the following instance of the destruction of eggs by these birds, and have no reason to doubt its authenticity:—

"Mr. Purves of Linton Burnfoot, near Kelso, had a tree on his farm in which a hooded crow had built her nest and hatched her eggs. Mr. Purves then went with the intention of destroying the young, but found he was too late, as they had flown. The ground around the tree was so thickly strewed with eggshells that he obtained the assistance of two friends and took the trouble to pick them up and count them, when they amounted to the large number of 196, all the eggs of the partridge, which had evidently been brought to feed the young."

The Rook (*Corvus frugilegus*), with perhaps one exception, is more numerous than any bird in our district; that exception is the jackdaw, which, though it does not assemble in immense flocks like the rook, yet, I think, equals it in numbers. Rookeries, great and small, are scattered all over our neighbourhood, those in Thoresby Park being the largest and most thickly populated. One of these, in a grove of Scotch fir and oak, about a quarter of a mile from the mansion, is of immense extent, and its occupants must be counted by thousands.

I have seen them in an evening when they were returning to their nests, quite darken the air with their flight, and on one or two occasions, when the turf has been infested more than usually with the larvæ of the cockchafer, they have literally blackened a patch of ground about a quarter of a mile square; and never shall I forget the amazement with which a relative of mine, fresh from a town residence, gazed on their countless numbers.

Their partiality for the grub of the cockchafer is productive of the most beneficial results. But I have seen long patches of sward in the forests and parks so thoroughly and uniformly dug up in their search for them that it was greensward no longer; not a patch as large as the hand had escaped being uprooted, clearly showing the abundance of these destructive larvæ.

I do not agree with the opinion so commonly expressed, that the bare space around the base of the bill of the rook is produced by its habit of grubbing in the ground; I have watched them very closely when they have been engaged in upturning the turf as I have described, and never saw the bill plunged beyond its length. Even when they are searching the newly-ploughed ground, I never observed any action which could produce the abraded appearance. I admit it is very natural to attribute it to such a cause; but is it not a singular fact, telling strongly against this theory, that in the extent of this bare skin there should be no appreciable difference in one bird over another, but all are equally denuded? Surely, if it was produced by digging, some variation in this would be noticeable, but I never saw such; the jackdaw, too, is as great a digger as the rook, and has a shorter bill, and yet the base is clothed with feathers which bear no trace of injury from such a cause. The editor of *The Field* favoured me with the following note on this point:—"There was a long discussion on this subject in *The Field* some years since, and several instances were mentioned in which rooks kept in confinement, where they could not dig, nevertheless lost the feathers. We also received the head of a rook in which the feathers were only partially removed; and those which yet remained were nearest to the point of the bill, and consequently, it may be supposed, would have been the first to suffer in the digging process, while those which had disappeared could scarcely have been removed by abrasion without injury to the feathers, which still existed in an untouched state."

Though naturally insect feeders, yet there are times when, pressed by hunger, rooks levy their contributions on the newly-springing corn, and in hard winters they will even frequent stackyards. They are very partial to potatoes, at least they are much addicted to digging up and carrying off those freshly planted, but it is chiefly at the time when their young are clamorous for food, " when there is little to earn and many to keep;" indeed they often suffer greatly from want at this time of the year. Macgillivray doubts the assertion that the rook pilfers freshly-planted potato sets, but I have seen them do so hundreds of times.

Though in our neighbourhood the corn is always tended by boys from the time of sowing until it is well out of the ground, in order to drive off the rooks, who would otherwise commit great havoc, yet I think the cultivators of the land have a pretty correct idea that, on the whole, the labours of these birds are productive of great benefit to the crops, and no greater destruction is made than of an occasional one, who, with wings extended by two split sticks, is placed *in terrorem* in the centre of a corn or potato field; and a very effectual scarecrow he makes—his constrained attitude is understood at a glance by his wary brethren, and they need no other hint. In some parts of the country the agriculturists are not so conversant with the habits of the rook, and I know that in one locality in an eastern county a large rookery was destroyed under the belief of the farmers that its inhabitants were hostile to their interests, and consumed a large quantity of corn. But mark the result. Two years passed away, and the farmers congratulated themselves on being rid of their winged foes, little thinking that they had other foes in their place whose approach was more difficult to detect. In the second year many fields of wheat suffered from wireworm; but in the third their ravages had become so general throughout the district as to occasion serious alarm. Little could be done to suppress their numbers until the rooks were again thought of, and the evil was traced to its true source. The rookery was permitted to be re-established by the return of many who had escaped the massacre, and who still cherished a partiality for their native trees, but who had hitherto been continually driven off. Their rapidly increasing numbers soon reduced the insect pest, leading the farmers to acknowledge the error into which they had fallen, and henceforth to look upon the rook as a friend instead of an enemy.

When rooks are feeding they always station several of their number as sentinels, and very faithful they are in sounding the alarm on the approach of a foe; they are not only vigilant in their watch, but evince a large amount of sagacity, an amusing instance of which was communicated to me by a friend on whose statement I can rely, and who witnessed the occurrence.

A very large field had been sown with wheat, and in the centre a little hut had been erected to shelter the boy who had to tend the field, and to enable him to reach all parts of it. A gentleman who wished to obtain a few birds to hang up in his own fields thought this would be a good opportunity of procuring them, for they thronged around in great numbers, and kept the boy actively employed to drive them off. So taking his gun, he went into the hut accompanied by the boy, and through some holes in the sides prepared to pour a volley on the invaders. But he reckoned without his host. The watchful sentinels seemed instinctively to divine the plot, their warning "caw" was loudly uttered, and the presence of the ambushed foe made known. They circled round and round and settled in the surrounding fields, but not one of them would trust himself within gunshot of the hut. For some time the gentleman waited in vain, and then sent the boy away with directions to walk straight out of the field; but this ruse did not succeed. The rooks still refused to "come and be killed," so he left the hut and followed the boy, but no sooner had he gone out of the gate of the field than the sentinels gave the signal, and scores of their fellows at once descended and commenced their foray. The sportsman determined not to be outwitted in this way, so he immediately took two persons with him into the hut and resumed his ambush, the rooks having taken flight on his reappearance. After a short time had elapsed he sent one of the persons away; and after another interval the second, expecting that as soon as they both left the field the rooks would return; but he was again doomed to disappointment; "beware" cawed the sentinels in the most sonorous tones, and none ventured to disregard the warning. Determined still further to test their powers of numeration, he again left the hut and returned with three persons, all four entering together. Again, one by one, the companions were sent away, and the plan was at last crowned with success; the rooks could count as far as three, but four was beyond their powers, and no sooner had the third person left the field than they hurried to the spoil, but only, alas! to leave two of their number dead on the field, victims to the want of a knowledge of numeration.*

I met with an interesting account of the sagacity of the rook in the *Dundee Courier* a few years since. Its truthfulness was vouched for by the gentleman who communicated it, and by the editor :—

"On Saturday week a very curious scene occurred in the colony of crows on the South Inch, Perth. One of the black denizens had been laboriously occupied in conveying sticks from the opposite side of the river, wherewith to build his nest, when something seemed to strike him that he was making no progress in its erection, and that he was the victim of some thievish neighbour. That his suspicions were correct he soon discovered, and evidently adopted the following plan to detect the culprit. He set off apparently to cross the river, and kept his usual way, but on reaching the island he suddenly wheeled round, and sweeping behind the lime sheds he reached his nest just in time to catch the suspected rogue in the very act of robbing him of a stick. A fierce engagement ensued, lasting several minutes, when the thief clearly having the worst of the fight, was compelled to render justice to his injured neighbour by restoring his stolen property, as for nearly half an hour after, the latter was seen to carry stick after stick from the other's nest without any molestation, and apply them to his own."

---

* A similar instance is given by Macgillivray of the carrion crow, from an account communicated to him by Mr. Weir, but in this case the crow proved a worse arithmetician than the rooks I have mentioned.

I can quite give credit to this anecdote, for I have known two similar cases in which one rook was detected stealing sticks from another; in both instances, however, the punishment was inflicted by more than the injured bird, and in one case with such severity that the offender's life was forfeited. I have more than once seen a rook chased from a rookery by a number of its inhabitants, but whether the hostility was shown because he was a stranger or a criminal I could not discover, but most likely the latter.

Lord Campbell, in his Lives of the Chancellors, says that "in Scotland the crows, who take such good care to keep out of gunshot on every 'lawful day,' on the Sabbath come close up to the houses, and seek their food within a few yards of the farmer and his men, discovering the occurrence of the sacred day from the ringing the bells and the discontinuance of labour in the fields, and knowing that while it lasts they are safe."

Various instances have been recorded of rooks eating eggs, and I once saw a pair on the 4th of June actively engaged for some time in chasing a pair of green plovers in a field on the verge of the forest. They were evidently bent on driving them away from a particular spot, which the plovers seemed as determined not to leave, and from their pertinacity I concluded that their nest was thereabouts, and that they suspected the rooks of a wish to plunder it, a conclusion which was no doubt correct.

The rook is occasionally subject to variations of plumage, and the saying of "as black as a crow" is not always applicable. In March, 1860, one was killed near us which was uniformly speckled with white.

I have already said that I think the Jackdaw (*C.*

monedula) is as numerous in our district as the rook, though it does not assemble in those large flocks in which the latter is seen. Its chief nesting places with us are not buildings, ruins, or cliffs, but the huge oaks which are the ornaments of our forest and parks. Every one of these ancient trees is more or less hollow, and two or three pairs, or even more, will make their abode in one tree; some of the cavities are very large, extending a great distance into the trunk of the tree, although the entrance may be only large enough to admit the bird.

When a hollow of this kind is selected it is astonishing to see what an immense mass of sticks is carried in for the purpose of raising the foundation to within a moderate distance of the entrance. I have seen cavities six or eight feet deep crammed with such a quantity of small sticks as would fill several wheelbarrows; and I have heard of an instance in which a small spiral stair in a church tower, which was seldom used, was so choked up with a similar accumulation, that when the door was opened no entrance could be effected until a quantity of sticks, sufficient to fill a cart, had been removed. In their strongholds in these hollow trees they rear their young in safety, and as comparatively few attacks are made upon them their numbers are very large. They are as pertinacious in their forays on the newly-sown corn as their larger brethren, the rooks, but their general labours are equally beneficial to the husbandman, larvæ being their chief and favourite food.

They are active, lively birds, and possess a large amount of cunning as well as impudence. I have seen them rob the dinner-baskets of the labourers in the fields; and it was most amusing to watch the stealthy, wary manner in which they effected their plunder—proving themselves most accomplished thieves.

They are sociable and friendly amongst themselves, and live in goodwill and peace towards their neighbours. I never saw amongst them any of those violent, ill-natured attacks which the rooks make upon some unfortunate individual who may not happen to belong to their coterie, but they appear quietly to do as they would be done by.

They mingle freely with the rooks when feeding, and are as active as they in their search for the larvæ of the cockchafer, digging up the turf with great perseverance. Why, then, is not the skin around the base of the bill as bare as the rook's? I have watched them most closely, but I never saw the slightest abrasion of their feathers, which must have been the case if produced by digging.

In confinement they manifest great familiarity, and are much attached to their owners, sometimes exhibiting a quaint comicality of manners which is very amusing, and greatly delighting in a bit of mischief; they will sit on the rail of a cottage garden watching the play of the children, and at dinner-time keeping a good look-out for their share.

The artfulness and thievish propensities of the crow family seem to be concentrated in the Magpie (*C. pica*). Wary to an extreme, it is ever ready for plunder, and, though often kept in a cage, I know of few common birds with whose general habits in a state of nature we are so little familiar. It is rarely that it permits of a near approach, except under favourable circumstances for concealment, but in our secluded districts I have often enjoyed these opportunities, and have been much interested in observing its lively habits. It is very fond of mingling with sheep, especially when feeding on turnips, and under cover of a hedge I have frequently stolen up and enjoyed a laugh at them. They search about for insects, now with a long elastic bound snapping a tick from a sheep's fleece; now looking up in its face with the utmost pertness, as much as to say, "I should like a peck at your eyes;" and then, with a few vigorous hops, away to another.

With regard to the haunts of the magpie, it appears most decidedly to prefer the cultivated farm land to the wilder forest, being rarely seen in the latter localities. In woods or plantations I never met with it; and its nest, as far as I have observed, is almost always placed in hedgerow trees. The ash appears to be more frequently chosen than any other. I have often admired the architectural beauty of the magpie's nest, though why it builds it with a dome it is difficult to say. Certainly the structure is too open to afford any protection from the weather, but at the same time, as the nest is generally placed in isolated trees, the dome may be designed to screen the eggs from a passing plunderer, and for that it is quite sufficient.

The assertions of some of our older naturalists that the magpie builds her nest with two entrances seems to want verification. I never met with one so contrived. Nevertheless, such a construction is followed by some birds, the pheasant cuckoo of Australia for instance (*Centropus phasianus*), which I know builds a domed nest with an opening on each side, from which the head and tail of the female project when she is sitting; it is therefore possible that there may be some truth in the story.

The appearance of the magpie used to be considered as an omen of varied significance according to the numbers seen. This superstition has greatly died out of late years, but a rhyme which is still common amongst us, and which I have known from a boy, records the popular belief as follows:—

> "One for sorrow,
> And two for mirth,
> Three for a wedding,
> And four for a birth;
> Five for a fiddle,
> And six for a dance,
> Seven for Old England,
> And eight for France!"

The meaning of the last four lines is not very apparent; perhaps the poet thought his stanza required a finish!

The Jay (*Garrulus glandarius*) is one of the most beautiful of our native birds; but he bears a bad character, from his predatory habits, and suffers accordingly. The keepers shoot every one they meet with, and one cannot go far in our woods without seeing their dead bodies dangling from the lower branches of a tree, and bleaching in the wind.

They are lively, restless birds, ever on the watch and ready to give the alarm with their harsh cry, whether it be quadruped or biped that appears. When it is possible to get near them unperceived (which is rarely the case), it is very interesting to watch their quick, active motions, the rapid raising and lowering of their crest as any other bird flies past, and the inquisitive glance of their bright blue eye; the ear too will be saluted with varied but not very musical sounds, their own natural harsh "wrake, wrake," or an imitation of the cries of some of their neighbours. Indeed, in their general habits they are very similar to their American cousins, described so inimitably by Alexander Wilson.

Jays seldom congregate in larger numbers than the brood of the year, these small parties of five or six generally associating together through the winter, and dispersing in the spring. Their flight is not extended far, and is of a broken, undulatory character, as they pass from tree to tree, or from one plantation to another. A high hedge or bush is generally chosen for the nest, which is constructed of dried sticks, the base being of larger ones, supporting a shallow cup, which is loosely woven of small twigs, and lined with fibrous roots. I once took one containing four eggs from the extreme top of a tall beech tree, at least fifty or sixty feet high; this was entirely made of dead birch twigs, with the exception of the lining of roots, and was much more neatly put together than usual. The late Mr. Waterton remarked that "the nest of the jay is *never* seen near the tops of trees;" the instance I have mentioned was certainly an exception to this rule, if it is one, which I greatly doubt.

The Woodpeckers are peculiarly inhabitants of the forest, and that handsome species the Green Woodpecker (*Picus viridis*), is very abundant, our old decaying oaks being a favourite resort, and furnishing them with an ample supply of food. The light, sandy forest soil is greatly frequented by ants, and here you are sure to meet with the green woodpecker. It is a shy species, and its white eye wears a peculiarly wild expression, while its singular cry, heard in the depths of the woods, has something very unearthly and startling about it. It is generally uttered while it is on the wing, making its odd, festooning flight from tree to tree, but not invariably so, as I have heard it both when clinging to a tree and when on the ground. The latter situation is only frequented where there are anthills, when it willingly leaves its strongholds the trees, to search for its favourite food.

Its motions on the trees, for which it is so admirably fitted, are well worth watching. I never saw it by any chance perch on the upper side of a bough, but it is fond of clinging to the under side, where during the day insects chiefly congregate. It is on the perpendicular trunk, however, that it is most at home. Commencing at the base it pursues a spiral course to the top, prying into every chink and crevice, tapping here and there with vigorous and rapid strokes to alarm its insect prey.

I have remarked previously that nearly all the old oaks in the forest have suffered the loss of their tops by the agency of wind and lightning, aided by natural decay. Sometimes you may see the upper portion of one of these venerable trunks quite denuded of its bark, and riven with many fissures, though the tree is all the while in vigorous growth. On some of these I have often noticed the green woodpecker practise a singular feat. Placing its bill in one of the long cracks I have mentioned, it produces, by an exceedingly rapid vibratory motion, a loud crashing noise, as if the tree was violently rent from top to bottom. I have heard it when the sound was so loud and sudden that the woods rang again. For a long time I was at a loss to know how it was produced, but I one day witnessed the process, and have seen it several times since. It would effectually rouse up all the insects, for it seemed as if the tree quivered from top to bottom.

Montagu mentions the jarring sound made by this species, but imagines it to be the call of both sexes to each other. With this I do not agree, but think from frequent observation that it is produced in the way I have mentioned for the purpose of procuring food.

The hole in which the eggs are laid is generally with us hewn through the sound outer portion of the trunk, until at a few inches deep the decaying wood is reached, in which the hollow for the eggs is formed, for nest there is none. I have met with one or two holes where the bird has evidently erred in its calculations. One in particular was about fifty feet from the ground, and had been begun in a tree too sound for the purpose; the hole was chiselled out of the solid wood, and must have cost its maker great labour, having been driven forward in a horizontal direction for about nine inches, but the wood continuing sound the bird had apparently become disheartened in her work, and abandoned it. When I first discovered it, it had not been long deserted, for I took the trouble to climb up and carefully examine it, measuring the depth with my stick, and ascertaining by the sound that the wood at the bottom was free from decay. I could not help wondering how the bird, in a hole not larger than the diameter of its own body, could find room to give those violent strokes with its bill which would be necessary to penetrate the solid oak.

The great Spotted Woodpecker (*P. major*), though by no means so plentiful as the last species, is still sufficiently so to prevent its being considered rare. It is an active climber, generally taking a diagonal course up a tree, and I have sometimes observed that when it has reached the base of a large arm, it has left the trunk, and with great rapidity run round the arm spirally for several yards of its length, and then flown off to the trunk, and resumed its course upwards. This motion, which was performed without a break, had a very singular appearance.

Thoresby Park is a favourite habitat of this species; there, from a large crab tree, I once roused three together; they uttered a short, sharp cry, as they usually do when disturbed, and flew off to a clump of large Scotch firs at a little distance—two of them, by the red occiput, being evidently males. In that part of the park known as the Old Wood, I have met with them more frequently than in any other, but seldom with more than one at a time; I have also seen them in Birkland. They are not such exclusive insect feeders as the preceding species, but vary their diet with the seeds of various trees, especially those of the pine.

The lesser Spotted Woodpecker (*P. minor*) is, I think, more local than any of our other British species. I have only met with three specimens—one a male, and the others two females; two were in Thoresby Park and one in Birkland; none of them had the shyness of the other two species. The male I watched especially for some time, while it was engaged in searching with most industrious agility the branches of a very large silver willow. It seemed little alarmed at my presence, but at length took wing to some trees at a short distance, repeating its cry several times in a shrill tone.

I have only met with the Wryneck (*Yunx torquilla*) a few times. It is well known by our country people as the "cuckoo's mate," but its shy and secluded habits remove it greatly from the common gaze. I have never seen it but in the forest, where it is attracted by the same inducements as the woodpeckers—abundance of ants, and hollow decaying trees. It is one of our most beautiful birds, though its charms do not consist of gay colours, but of minute and exquisitely varied pencillings which it is impossible to describe. I never found its nest but once, when three eggs occupied a shallow and much exposed cavity in a decayed oak tree. My attention was drawn to it by the female, which was perched on a bough of the tree, and which, after suddenly raising the feathers of her head, flew off to a short distance.

In every part of our wooded district the little Creeper (*Certhia familiaris*) finds a home. Summer and winter, if you watch carefully and quietly, a glimpse will be had of its little brown figure gliding up the trunk of some tree like a mouse, and if your person is concealed, you may see it prying with its slender bill into the crevices of the bark for spiders and other insects that lurk there; but the moment you are perceived it creeps round to the opposite side of the tree, or flits to another at a little distance. Its chirp is very weak and humble in tone, as if it was afraid of being noticed, and yet in the summer time it may be heard oftener than it can be seen. Indeed, so retiring and unobtrusive are its habits altogether, that a careless observer might fail to see it at all.

In the winter I have noticed it frequenting barns and other outbuildings, and the neighbourhood of houses, the warmth of which attracts a large number of insects; I have also seen it searching the fences in my garden. At such times it loses somewhat of its usual timidity, although it is still very shy.

Of the Common Wren (*Troglodytes Europæus*) it is hardly necessary to say more than that it is a most familiar and abundant species. Every child knows and delights to see "little Jenny Wren," the very picture of vigorous, bustling industry and pert independence; and its share in the tragic story of " Who killed Cock Robin ?" will ever make it familiar to our children.

It is always interesting to watch the active vagaries of these birds as they half flit, half creep in the bushes and hedges of our gardens; they are bold little creatures, approaching within a yard or two without fear, but at the same time vigilantly alive to secure their own safety. While I write there is one in a barberry bush just outside my window, so busy and bustling in its activity, and with its tail cocked up at right angles with such a consequential air, as fairly to provoke a burst of laughter from my children.

I hardly know any bird that employs such various materials in the construction of its nest as the wren. Moss is the most generally used, but it seems to avail itself of those substances which lie most conveniently for use, and these are often selected with an evident view to concealment, or at least that end is attained, whether designedly or not. On the other hand, sites for the nest are frequently chosen in the most public situations, as though privacy was scorned; but these are exceptions.

It has often been noticed as a singular circumstance that so many unfinished nests of the wren should be found, and one year I counted six at one time in the creepers outside a summer-house or rustic temple in the pleasure grounds at Thoresby, within the space of a few yards. They were in various stages of construction, though none of them were completed, but seemed to have been abandoned one after the other. A writer in Loudon's Magazine of Natural History, vol. iii. p. 568, broached the theory that the male bird, from want of having nothing better to do during the incubation of the female, keeps his hand—I beg pardon, his bill—in practice by constructing "cock nests!" This idea, however, seems to me to be not very probable. A likelier cause, perhaps, is that the wren may be more fastidious than other birds, and suffers itself to be affected by very slight causes of disturbance, and so after a nest has been partially constructed it deserts it and commences another, and this several times in succession. This, I think, is far more probable than the erection of "cock nests" by the male. Mr. Neville Wood says that the wren "often builds itself a dwelling in autumn, and lodges in it on cold nights." Mr. Weir states the same, and both are thus quoted by Macgillivray, who is of the like opinion. The nests I found, as mentioned above, were in spring, and were recent erections; and though it may have a habit of erecting nests for winter roosting places, yet I scarcely think these would be commenced so early in the year.

I am pleased to be able to record a single instance of the Hoopoe (*Upupa epops*) visiting our district. A male in fine plumage was shot on the forest a few miles north of Ollerton, but I know of no other occurrence of this rare and handsome bird.

The pretty and chastely-coloured Nuthatch (*Sitta Europæa*), though it is rather locally distributed in England, is by no means rare with us; indeed, in some places it is plentiful. The large kitchen gardens at Thoresby, which stand in the midst of the park, are a very favourite haunt, the attraction being a long row of large and aged nut trees which skirts the southern side. There I have often watched their busy operations in nutting time. The nuts are of various kinds; the filberts are made short work of, but several of the trees bear a large cob nut with a very thick shell, and into these they are sometimes puzzled to find an entrance. Two of the posts in the garden-fence were constantly resorted to in consequence of their being split, and in these cracks they fixed the nuts with great dexterity, and were thus enabled to break them with ease. A slight cavity in a fork of one of the trees was also used for the same purpose, and their loud hammering might be heard for a considerable distance.

It is only on a tree that they are seen to full advantage; there they are perfectly at home; up or down the trunk they glide with equal facility, and rarely resort to the ground. I have seen them do so to pick up a nut they had let fall, but they appeared to move awkwardly on a flat surface, and flew back to the tree the moment the nut was secured.

The nest of the nuthatch, if it can be called a nest, is always placed in a hollow tree, and is generally constructed of dried leaves or moss very carelessly deposited. I took the eggs from one in 1854, which was composed of dry grass. The five eggs it contained were of the usual white, marked with brown; but in this instance they exhibited a singular gradation of colour, the egg which had apparently been first laid having the markings dark and numerous, each one of the others being less so, until the one which I consider was last deposited had only a few minute specks of pale brown, the glands which secrete the colouring matter having evidently become exhausted. I have remarked this gradation in colour in the eggs of other species. Those I have just mentioned were taken out of a hollow in a decayed oak tree, the entrance to which was only about six feet from the ground, and (as usual where the opening is too large) was contracted by a plastering of clay.

Few of our native birds possess such pleasing associations as the Cuckoo (*Cuculus canorus*). The very name carries us back to the times of merry childhood, and recalls the feeling of joy which the monotonous though musical note awakened in our breasts. We knew that winter was gone, and that violets and primroses were to be found in the woods. How blithely did we set off to gather them, and how pleased were we to imitate the well-known call of the "harbinger of spring!"

Yet it is only its association with the joyous springtime that makes us welcome the cuckoo, for the bird itself possesses none of those attractive qualities which naturally call forth our admiration. It elicits in us no sympathy, for it exhibits no fidelity to its mate, no affection or tender solicitude for its offspring, but, scattered here and there, it leaves its young ones to the protection and care of strangers. Yet we cannot blame it for this, as it does but obey its natural instincts. It has often struck me as one of the many marvellous ways which our Divine Creator has devised for the preservation of species, that the foster parents never seem to discover the fraud perpetrated upon them, but hatch the strange egg and tend their foundling with as much care as their own offspring.

Of few of our British birds have such various assertions and opinions been hazarded as of the cuckoo, some no doubt arising from want of observation, others from observations carelessly made. Some have stated that the cuckoo has been known to feed her own young one; this has been denied by others, who have asserted that in the alleged instances the young of the goatsucker has been mistaken for that of the cuckoo—a mistake which might easily be made, from the similarity of the plumage of the young of both species.

The mode in which the cuckoo introduces her egg into the nest of another bird has been made a greater difficulty of than necessary, from the fact that usually it would be a physical impossibility for the cuckoo to enter the nest and lay her egg in the ordinary manner. I have often found the egg, but never, (except in one instance, where the shallow nest of the pied wagtail was chosen), was the nest in such a position as to be reached by the cuckoo otherwise than with her bill. The latter is doubtless the instrument by which the egg is deposited in its chosen place.

But of all the extraordinary theories which have been brought forward respecting the cuckoo, that advanced by Dr. Baldamus of Stuttgart is the most amazing. It was first published by him in 1853 in the *Naumannia*, the leading ornithological periodical of Germany, but had remained unknown to English naturalists until the Rev. A. C. Smith called attention to it in the *Zoologist* for March, 1868, and gave a translation of Dr. Baldamus's paper in that periodical for the following month. I would advise all my readers who are interested in the subject to peruse that article for themselves, but for those who have no opportunity of doing so I will give a brief outline of the theory. Dr. Baldamus begins by asserting that the eggs of the cuckoo are subject to great variation, both in colour and markings, and that he had found thirty-seven varieties! He then set himself to discover the cause of this singular variability, and after some time spent in diligent research and examination, he communicated his views in the article in question. Dr. Baldamus says he discovered that the cuckoo deposits her eggs in the nests of thirty-seven different species, and that in by far the greater number of instances these eggs bore the same colour and markings as the eggs of the birds in whose nests they were laid. He enters into details, and proves this to his entire satisfaction, giving a list of all the species in whose nests he and his friends found cuckoo's eggs, summing up the question thus:—"Therefore I do not hesitate to set forth as a law of nature, that *the eggs of the cuckoo are, in a very considerable degree, coloured and marked like the eggs of those birds in whose nests they are about to be laid, in order that they might the less easily be recognised by the foster parents as substituted ones.*"

He then asks the question, "*Does the same hen cuckoo lay eggs of the same colour and markings only? and so, is she limited to the nests of but one species? Or else, does the same individual lay eggs of different colour and markings, according to the character of the eggs amongst which her own will be intruded?*"

In discussing these points the Doctor considers it by no means "improbable" "that the sight of the eggs lying in the nest has such an influence on the hen which is just about to lay, that the egg which is ready to be laid assumes the colour and markings of those before her," and adduces as evidence in proof the account of the proceeding of the patriarch Jacob as given in Genesis xxx. 37, &c.

His final conclusion, however, is this, "*that every hen cuckoo lays only eggs of one colouring, and con-*

sequently (as a general rule) *lays only in the nests of one species.*" Such is, I believe, a fair outline of the theory, and the italics are the Doctor's, not mine.

The publication of the article in the *Zoologist* elicited a discussion in the *Field*, in which Mr. Hewitson, Mr. Newman, Dr. Bree, Mr. Dawson Rowley, and others took part, the weight of opinion being adverse to the theory. It is but just that the views of a man of eminence like Dr. Baldamus, made in good faith, however startling and contrary to our preconceived conclusions, should be received with respect, and be subjected to a careful examination. I must, however, frankly state that I cannot accept his conclusions, because I doubt the facts from which they are drawn.

I never found the eggs of the cuckoo to vary in any great degree; they most resemble in colour and markings the light varieties of the skylark's, and next, those of the pied wagtail, and these I believe to be the usual types. So eminent and experienced an oologist as Mr. Hewitson holds this opinion, and Mr. Rowley, who possesses the nests and eggs of fourteen species, with a cuckoo's egg in each, twelve of which he took with his own hands, says that out of all these, only two (and those the two he did not take himself) display any resemblance between the intruder and its companions, and in both to him it is very faint. I believe that what Dr. Baldamus supposed were cuckoo's eggs, were only abnormally large ones of the birds in whose nests they were found, and this variation in size is well known to every oologist and every bird's-nesting schoolboy. It would be strange indeed if the cuckoos laid these varied eggs in Germany and not in England, and yet British ornithologists have never discovered them. Another point on which the Doctor lays stress is what he calls the *grain* of the shell. I have been unable to detect this. I have examined under a low magnifying power the eggs of the cuckoo, as well as those of the hedge-sparrow, the meadow pipit, the pied wagtail, and others, and can see no difference in this respect.

I am at a loss to see what purpose can be served by such an alleged resemblance. The German professor says it is to prevent them being detected by the owners of the nest, and ejected or destroyed, and thus the continuance of the species is insured. Is this necessary? If so, the American Cow Bird (*Icterus pecoris*), which deposits its eggs in the same manner as the cuckoo, should possess this advantage for the same ends: and yet Wilson says of it, "these odd-looking eggs were all of the same colour, and marked nearly in the same manner, in whatever nest they lay, though frequently the eggs beside them were of a quite different tint." No variation is discoverable here, and yet the species does not fail; but, reasoning from analogy, if it is necessary in one case, it is in the other.

There are two conclusions to which we are shut up by this theory. First, every cuckoo must possess the power of colouring her eggs at will; or, secondly, there are thirty-seven kinds, each kind laying different eggs, but which are constant in their colour and markings.

With regard to the first, I believe a cuckoo does not seek a nest until her egg is ready for extrusion, and consequently mature, having received its colour and markings from the glands in the lower portion of the oviduct, and this is confirmed by the opinion held by many naturalists, that she possesses the power of retaining her egg after it is ready, until she can find a nest for its deposition, when the act is a very rapid one. How then could a sight of the eggs in a nest alter the colour of her egg already matured?

Dr. Baldamus finally gives up this conclusion in favour of the second, which I think is as unsupported as the first. Of course he admits that all the thirty-seven kinds of cuckoos are *specifically identical* as Cuculus canorus, and yet in one important point, the colour of the eggs they lay, he alleges they are *specifically distinct*. I cannot, I confess, understand such an anomaly. The thirty-seven kinds are identical in structure, plumage, and size, and yet each lays a differently marked egg. But are we to suppose that these thirty-seven kinds, visit a country at the same time, feed on the same food, mingle in the same hedgerows, and yet do not breed together? The idea is incredible, and yet the crossing which must of necessity take place between these imaginary kinds, would of course destroy in time the alleged distinctive markings of the eggs their progeny would produce, and bring all to one uniform character.

I have dwelt at length on this subject, which is one of great interest, but I must say that I should like clearer proof than I at present possess, before I can accept either conclusion. All who possess the opportunity should lend their aid in its investigation, and accumulate such evidence as will either disprove the theory, or make us willing to give up our long-cherished opinions.

The first point evidently is, *do* the eggs of the cuckoo vary to the extent asserted? I believe they do not; and to this our observations should be directed, and nothing but positive evidence admitted.

On the 20th of August, 1860, I witnessed with great interest a pied wagtail feeding a young cuckoo. I was crossing the bridge in the village, when I saw the cuckoo perched on the upper rail of a fence which divided the meadow from the stream, the spot where it sat being about fifteen yards below the bridge. The stream was shallow and partially filled with weed-beds, and on these the wagtail was running in its usual rapid manner, seizing first one insect and then another, which it directly conveyed to its foster-child on the fence. There the great overgrown baby sat, eagerly receiving the food from its tiny friend, but looking far more able to provide for itself. I stood on the bridge watching the pair for a quarter of an hour, and during the whole of this time the wagtail was constantly feeding the cuckoo, which sat so quietly that I thought it was unable to fly far, and that perhaps I might effect its capture. I accordingly got over the hedge into the meadow, and went cautiously towards the spot, which it allowed me to do until I was about three yards from it, when it flew off and settled on a pump that stood in the meadow at a short distance. The poor wagtail seemed distressed, and followed it to the pump, where it again resumed its feeding. On my approaching a little nearer it again took flight, but with such strength of wing as to convince me that I had been mistaken in thinking I could make it a prisoner. It settled on the top of an alder tree, and from there flew out of sight, the little wagtail faithfully following in its wake. It was evidently a strong, vigorous bird, equal to a long flight, and would doubtless soon take its departure.

Some have supposed that the cry of the cuckoo is only uttered by the male bird, but this has been denied by many others; I disbelieve it myself, for I have positive proof that the note is uttered by both sexes, from having shot the female when thus engaged. I have met with equally undeniable proof of its egg-sucking propensities, for a friend of mine shot one in a garden a short distance from my own, his attention having been drawn to it by the well-known cry. As he went into the garden the bird rose from the foot of the hedge, and was immediately brought down; when he picked it up it was not quite dead, and as he held it, it laid an egg in his hand, thus being another instance of the female uttering the cry. The bill of this bird was covered with yolk of egg, which was also spread over the feathers at the base. On proceeding to the spot from which it rose, the cause of this was at once seen; for there was the nest of a pied wagtail, with all the eggs broken. It seemed as if the cuckoo had greedily plunged her bill amongst them, and thus smeared the yolk over the feathers of her face.

The common cry is uttered both while on the wing and when perching; in the latter case the bird lowers and raises its head at each utterance, spreading out its tail, and partially swinging itself round as if on a pivot at the same time. It has also another note of a very liquid character, resembling the syllable "quille," which I have remarked it repeat quickly five or six times in succession, and generally after it has alighted on a tree.

The earliest arrival of the cuckoo I have noted was on Feb. 16, 1849. The weather was particularly fine and sunny for the time of the year, and the cry at that unusual season attracted the attention of a number of persons. It was heard for half an hour on that day, but not subsequently. Could this have been a late-hatched bird which had not migrated? I should hardly be inclined to entertain such a supposition; for where could it have found caterpillars, grasshoppers, and other summer insects, which are its chief food, to support it through the winter? I am more disposed to think it was an unusually early arrival, and that, as it was not heard again, it most likely perished from cold and want of food.

None of our birds can boast of more beautiful plumage than the little Kingfisher (*Alcedo ispida*); the glossy metallic blues and greens with which it is adorned seem to belong more to the parrots, trogons, and other species peculiar to warm or tropical countries. Indeed, when I watch the rapid flight of a kingfisher, it always reminds me strongly of some of the Australian parrakeets, especially the *Lathamus discolor*, the metallic colours of whose plumage show most brilliantly during their glancing flight, particularly when the sun is shining.

The kingfisher is constantly to be seen, and yet is not an abundant species with us. The two small streams, the Morn and the Idle, which intersect our forest district, are very favourable to its habits and requirements; but though it delights to seek its food in secluded spots, it does not confine itself to such, and I have repeatedly seen several glancing up and down the stream which runs past the village, and darting through the arches of the bridge as I stood on it. I have even taken its eggs from a hole in the bank of the stream within a stone-throw of some houses, and of my own garden.

The nest of the kingfisher is another of those questions on which naturalists have greatly differed, and I know not that my own observations have enabled me to throw much light on the matter. Whether a layer of fishbones is *purposely* laid for the reception of the eggs, as Montagu asserts, I cannot say, but in every nest I have examined I have never found any other material used, nor have I ever seen the eggs on the bare ground. In every instance they rested on a layer of the castings, which were slightly hollowed for them, though the latter form may have been produced by the mere weight of the parent bird while the eggs were laid. I believe that a deserted hole of the water-rat is usually chosen, and that they rarely excavate for themselves.

The young ones, after they have left the nest, are exceedingly clamorous; so much so, that their loud, shrill twitterings were once the cause of my witnessing the interesting scene of a brood being fed by their parents. They were six in number, and were perched on the boughs of a dead bush overhanging the stream. They seemed very voracious; for though both the parent birds were constantly bringing them food—sometimes a small fish, sometimes what appeared like a slug or leech—they apparently failed in satisfying their appetites, and every fresh supply was eagerly competed for, sometimes a sort of scuffle taking place as to which was to receive it. During the absence of the parents the young ones sat very quietly; but the distant approach of the old ones was quickly perceived, and in a moment their listless attitude was changed into one of animation, they stretched themselves eagerly forward, and with loud twitterings and open mouths showed how expectant they were.

The Chimney Swallow (*Hirundo rustica*) is almost as much a household bird as the robin; it appeals to our better nature by the fearless confidence with which it seeks our dwellings, as the sanctuary where its tender young will be safe; and this, combined with its gentle, pleasing manners, justly makes it a general favourite. Wherever the swallow is found it seems to possess the same instinctive confidence in man, and the same preference for buildings.

In this country a chimney is most generally chosen by the swallow wherein to erect its nest; but in this selection I have never observed it show any particular preference for a shaft in a stack of chimneys more than for an isolated one. I fancy the only condition which seems greatly to influence it in this respect is, that it shall not be one which is in constant use. In my father's house there was an isolated chimney, which certainly was not used more than once or twice a year, and for at least thirty years I never knew this without a nest. It was a short, straight shaft, up which when a boy I have often looked with longing eyes at the prize above; and once or twice I remember an unfortunate young one tumbling down into the empty fireplace when essaying to leave the nest on its first journey. There was a window at a short distance, nearly on a level with the chimney-top, and I have spent hours, at various times, in watching the busy labours of the parent birds in constructing and repairing their nest. In some years the winter rain and snow would be so heavy as to demolish the frail structure, when a new one had to be built; in others it merely required a little patching, or a new lining of feathers, to make it habitable; but, with very few exceptions, the same angle of the chimney was always selected for the new nest, and it never varied more than a few inches in its distance from the top.

Though the swallow does not rank high as a songster,

yet it has a very pleasing and melodious warble; it chiefly indulges in this early in the morning, even long before sunrise, or towards evening, and it is quite in keeping with the gentle character of the bird.

By what extraordinary instinct do the swallow and its congeners ascertain what weather is prevailing in this country, for such really seems to be the case? In some years, when the season has been backward, I have remarked a few pairs only arrive, and even these have seemed after a day or two to disappear. In the year 1849 the spring was particularly backward; April was cold and bleak, and unfavourable to the development of insect life; and not until the 11th of May did any of the hirundines make their appearance, on which day I first noted a few pairs of the common swallow and the house martin, but the main body did not arrive until three days afterwards. This was no local occurrence, for the same ungenial weather was general throughout England. In 1847 I noticed the same phenomenon, under precisely similar circumstances as regards the weather. A few pairs arrived on the 29th of April, but immediately departed, and I saw no more until the 4th of May, when I remarked a single pair of swallows; but these were not joined by the main body until the 6th.

By what mysterious system of telegraphy was the intelligence conveyed to the southern voyagers that their journey had better be delayed for a time? We boast of our wisdom and intelligence, but how little able are we to elucidate facts like these.

During the time of building I have often seen the swallows frequent the gutters, or any wet place in the village street, from whence they obtain the mud for their nests; their feet, however, and short legs do not well fit them for walking, and I never saw one make more than two or perhaps three steps without using its wings. As soon as the young ones can fly they are in the habit of resting all together on the branch of a tree, generally choosing a withered one, and here the parent bird feeds them as she passes on the wing.

The swallow is very vigilant to detect the presence of any bird of prey; one or two wild hurried shrieks are uttered by the first who becomes aware of the danger, the call to arms is immediately obeyed, and in a few seconds all within hearing of the note of alarm are gathered together, and fly wildly about their enemy. The cuckoo is pursued in this way quite as much as any of the hawks.

How strange it is that the idea that swallows wintered in the mud at the bottom of ponds and rivers should ever have been a matter of belief with intelligent and scientific men, and have been so long and pertinaciously held; and stranger still that in this boasted age of enlightenment the wild story seems to be yet believed. Only last year I met with a paragraph in a serial circulating entirely amongst the educated classes, which stated this as a fact about which there were not two opinions. I opened my eyes in astonishment to read such information as the following:—"In Sweden the swallows, as soon as the winter begins to approach, plunge themselves into the lakes, where they remain asleep and hide under the ice till the return of the summer, when revived by the new warmth, they come out and fly away as formerly. While the lakes are frozen, if somebody will break the ice in those parts where it appears darker than the rest, he will find masses of swallows—cold, asleep, and half dead; which, by taking out of their retreat and warming, either with his hands or before a fire, he will see gradually to vivify again and fly. In other countries they retire very often to the caverns under the rocks. As many of these exist between the city of Caen and the sea, on the banks of the Orne, there are found sometimes during the winter piles of swallows suspended in these vaults, like bundles of grapes. We have witnessed the same thing in Italy." How the supposition can be entertained that a hot-blooded and lung-breathing creature like a bird can undergo immersion in water for months, and not be drowned, passes my comprehension. A single experiment would at once have demonstrated the absurdity of the theory, and proved that a swallow is no more fitted to live under water than a man. That some swallows have been found during the winter in a dormant condition has often been proved. These are most likely late-hatched birds; but I think it is very questionable if they ever survive the winter in a torpid state, and when such have been accidentally disturbed and roused into temporary activity, they almost immediately disappear again, and doubtless perish from want of food. In January, 1842, I knew of an instance in which a pair of chimney swallows fluttered out of the thatch of an old barn which was being pulled down. They seemed in great distress, and after flying about the place during that and the following day, nothing more was seen of them.

The late mild winter seems to have led a few pairs of this species to remain with us to an unusually late period. Five or six were seen skimming about at Sark on the 26th October last, and two or three at Margate at the latter end of December last. All these showed great vigour and liveliness, very unlike the languid fluttering of the pair above mentioned.

The House Martin (*H. urbica*) is generally a few days later in its arrival than the swallow (whether the latter be early or not), the same atmospheric influences affecting both species equally in limiting or increasing the supply of food. Its flight is less powerful than that of the swallow, or perhaps it merely appears so from lacking all those vigorous swoops which mark the course of the latter; but it often flies with a wavering motion, as if uncertain of its destination.

The martin is, equally with the swallow, an attendant upon civilization, and loves to associate its dwellings with those of man; indeed, the situations it generally selects for its nest are such as to bring its nidification more immediately under our notice than that of any other of our native birds, and a pretty sight it is to watch their busy operations.

The eaves of buildings or the corners of windows are their most favourite spots; but I have never met with a nest in such places open at the top, as I have frequently seen it represented in works of natural history. In one recent book, the illustrations of which are generally very faithful, the nest is figured as a shallow dish fixed to a wall, and entirely open at the top. Surely this must be a mistake; or, if drawn from nature, it cannot be taken as the type of the nest of this species. All that I have ever seen have had their walls carried up until they met the projection under which they were built, leaving a rounded hole in the middle immediately under the angle of the tile or cornice.

I saw a very remarkable instance of variation from the ordinary situation of the nest of the martin in the summer of 1864. The nest was built of the usual materials, but was placed on the top batten of a door in the wall of the gardens at Rufford Abbey, the batten being very broad, and extending to within a few inches of the top of the door. The nest was not only attached firmly to the door, but also to the lintel, and when first discovered had to be cut away from the latter in order that the door might be opened. This, however, did not divert the intentions of the parent birds: the eggs were duly laid and duly hatched; and when I saw it the young birds were nearly ready for flight. The door was in constant use, being opened thirty or forty times a day; but this frequent and sometimes sudden motion seemed in no way to alarm either old or young, the latter suffering me to touch them.

The martin, like the swallow, is fond of frequenting the ruts and gutters of roads for the purpose of picking up the mud with which it builds, which, it is very evident, is rendered more retentive by being tempered with the saliva of the bird. In some districts, where the soil is a strong clay, as it is a few miles from Ollerton, little tempering may be needed; but in our own neighbourhood, where the soil is very light and sandy, more preparation of the kind would, we should think, be necessary; yet with materials so different the nests they build with us are not less strong than those they erect in the clay villages.

A house opposite my own has been resorted to year by year as long as I can remember. There, under the eaves of the tiled roof, five or six nests were always placed. Sometimes, on account of the annoyance occasioned by their excrement, the occupant of the house caused those to be broken down which were directly over the door or windows, the others being left unmolested. The latter were always retenanted in the spring after receiving needful repairs, and new ones would be built on the foundations of the old ones, or occasionally a fresh site would be selected.

In my notice of the sparrow I have mentioned their fondness for taking possession of a martin's nest during their temporary absence. The martins were always greatly distressed by the aggression, flying wildly to and fro, and by their cries of alarm bringing a large number of their fellows to the rescue; but no active measures were taken, they contenting themselves by incessantly flying up to the entrance of the nest and giving utterance to their strong indignation—which, as a faithful chronicler, I am bound to say appeared to be entirely disregarded by the sparrow. Rarely has a summer passed without this scene being repeated two or three times in the same group of nests; and being just opposite to my own windows, it afforded me a great fund of amusement.

In 1835 a pair of martins built their nest under the eaves of a house at Sutton, a village a few miles distant; but during a violent thunderstorm it was partly demolished, and two young unfledged birds fell from it to the ground, but were apparently uninjured. The owner of the house, with great humanity, directed one of his men to procure a ladder; a board was placed under the nest, and secured by a couple of iron holdfasts, and the nest was then repaired with clay as well as it could be, a little cotton wool added to make good the lining. When this was done, the two young ones were replaced in the nest; but the most extraordinary part was that the parent bird, which was in the nest at the time part of it fell, remained sitting in the uninjured portion during the whole time it was being repaired, without exhibiting any alarm at the unusual proceedings. When all was finished, and her young ones restored, she flew around for some minutes, chirping cheerfully all the time, as if expressing her thanks for the kindness shown her.

The martin is much infested with a very disgusting-looking insect, as large, or larger, than the common bug. I have seen them swarm so thickly on some that the birds were rendered quite incapable of flight. I picked up one in this condition from the gravel walk in my garden. The poor thing manifested no alarm; but a glance told me the reason of this, as the bugs were creeping in and out of its feathers in numbers. I took it to the stream at the bottom of my garden, and got rid of its tormentors in the same way as the fox is reported to do. It seemed really grateful for the assistance, and as soon as the operation was completed flew off with the greatest alacrity.

The next species, the Sand Martin (*H. riparia*), is, if possible, more subject still to these insect pests, and it is rather singular that two years after the occurrence I have just mentioned, I found in my garden a sand martin in a precisely similar condition, and freed it in the same manner. I have sometimes found their nests abound with fleas, and this was particularly the case with a group of some twenty nests which had been excavated in the sides of a shallow gravel pit in the forest. The pit was not more than five feet deep on its steepest side, and the holes were formed in a stratum of slightly hardened sand, which was only ten or twelve inches beneath the heather-covered surface of the ground. At least a third of these had been abandoned;

for, wishing to satisfy myself on several points of their economy, I took the opportunity of the nests being within easy reach to examine them. All the empty ones, with the exception of one, swarmed with fleas, and I have no doubt had been forsaken from that cause.

This little swallow generally arrives from a week to a fortnight before either of the others; in some instances I have known it make its first appearance in March, quite regardless of the bleak cold winds. It is not, with us at least, seen much in the neighbourhood of houses, but prefers to seek its food over the meadows, and especially over water; the stream that flows through the outskirts of the village is always frequented, while I have rarely seen them in the street.

The flight of the sand martin, though rapid, is much less powerful than that of either the swallow or the martin, and I rarely remember to have seen it at those great altitudes attained by the latter in fine weather; it appears to prefer skimming just above the surface of either stream or meadow.

It is a pretty sight to watch a sandbank where their nests are abundant at the time they have young ones. Both parents take their part in feeding them, and during the greater part of the day it is seldom that many seconds elapse without one or other of them arriving with a supply. They are notwithstanding very discursive, and I have seen them constantly hawking for flies between two and three miles from their nearest nests.

The Swift (*Cypselus apus*), though found in all parts of England, is very variable in regard to the numbers frequenting any particular locality. In some I know, two or three pairs are the most I have seen, while in our own district it is abundant; but nowhere, I believe, does it equal in numbers either of the other hirundines. They are of social habit, delighting to hunt in company, generally in parties of from two to eight or ten pairs; and few sights are more beautiful than to watch a flock as they dart past like an arrow, or wheel with impetuous flight round the church tower, *squealing* all the time as if in the very exuberance of unfettered liberty and joy; and unfettered it truly is, for in the power, rapidity, and elegance of its ordinary flight, and its untiring activity, it has few rivals and still fewer equals.

The swift's period of residence with us is a very brief one, rarely exceeding two months and a half. No avant-courier precedes their arrival, which is sudden and simultaneous; one day none are to be seen, and the next they are in full force, making the air ring with their shrill cry. Their departure is also equally sudden, though not always so simultaneous, as now and then you may meet with a pair detained, perhaps for a week or two after their friends have departed, by domestic cares. The first week in May is the usual time of their appearance; but in 1854 I remarked an exception to the general rule, a single bird making its appearance on the 4th of April. This is the earliest date, and the 15th of May the latest, on which I have noted their arrival.

Some writers say the swift sometimes lays three or four eggs, but I never met with more than two. The old ducal mansions in our neighbourhood are always tenanted by several pairs, and the roofs of old houses are also frequented, the birds gaining entrance under the pantiles. These last are the most favourite resorts of the house sparrow, whose nest, I believe, is very frequently made use of by the swift without the owner's permission. The hollow limbs of the old oaks in the forest are also chosen, though but very seldom.

The heathy character of our forests and parks is very favourably adapted to the habits and requirements of the Nightjar (*Caprimulgus Europœus*), and during its brief visit it is here numerously distributed. Its common appellation is the fern-owl, for its jarring note is well known to every one; but yet few are familiar with its form, or would recognise in the day, the bird which they have chiefly seen darkly glancing in the dim and fading twilight, or perhaps only heard. It seldom makes its appearance in the daytime, but even then it does not exhibit that half sleepy, half stupid character which some of our other nocturnal birds do. I have met with it when it has been perched in its usual position lengthwise on a bough; when basking on a grassy bank in the sunshine; and once when busily engaged in half burying itself in a patch of loose dry sand; and on each occasion, although it allowed me to approach within a short distance, it showed itself quite awake to its own safety. At one time I attributed this fearlessness to stupidity, and meeting one day with one sitting on a grassy ride in the forest, it allowed me to approach so closely that I flattered myself I could effect its capture, and accordingly pounced on it with hat in hand. It took wing in a moment, settling again about a dozen yards further on, when I repeated my experiment with a like result, and then came to the conclusion that I was the more stupid of the two.

The nightjar is one of our latest visitors, generally arriving about the middle of May, though, like many other of our migratory birds, its arrival is hastened or retarded by atmospheric causes. I have seen them as early as the 21st of April (1847), and in 1853 I neither heard nor saw one until the 4th of June; this was very late indeed, but it was a most ungenial season, as may be gathered from the fact that on the night of the 13th of May there were four degrees of frost, and so cold and backward did the weather continue that there was little grass in the meadows on the 1st of June.

The flight of the nightjar is very light and buoyant, and almost as noiseless as that of the owls; though, from the nature of its food, the necessity for the latter quality is not apparent. When hawking for food it glides in graceful circles round the trees, every now and then doubling on its course in the most rapid and sudden manner. From close observation, I am of opinion that these abrupt turns are not mere capricious changes in its line of flight, but are occasioned by its making a dart at a moth or chafer. I have repeatedly tested this by throwing up a small stone as the bird flew over my head, when it would invariably make a plunge at it in the way I have described.

After the young are hatched, the parent birds are very watchful against any approach to them; and when walking in the forest in the evening I have constantly had them swoop at my head in a threatening manner, and sometimes so closely as to touch my hat with their wings. During the day the female rarely leaves either her eggs or young, and if disturbed feigns lameness, in the manner of the partridge, to draw off her enemy.

The use of the serrated claw of the goatsucker has been, and still is, a disputed question. I have watched the birds closely in a district where they are particularly plentiful, and have spent much time in carefully endeavouring to discover the purpose for which this claw is used. White's idea that the bird captures insects with its foot, and that the toothed claw is to give it a firmer grasp, is a pretty one, though I do not think this is its usual practice; if it is, we might suppose that a careful observer could not fail in time to mark the action of the leg as it struck at its prey; but during many years' attentive observation I have failed in seeing it used in this way. Bishop Stanley is of White's opinion, and says that its singular habit of "dropping or tumbling over as if shot" is in consequence of the bird losing its balance as it puts its food into its mouth with its foot. I believe, however, an examination of the claw itself would be against White's theory. It is slightly flattened and curved outwards, and it is the inner or convex edge that is pectinated, *not* the under side, which would make it the most effective instrument for grasping. Then the kestrel, which is known to catch cockchafers with its foot, should also have a serrated claw, but it is entirely devoid of anything of the kind.

Alexander Wilson's opinion is that in the American species it is employed for the purpose of freeing the plumage of the head from vermin, that, he says, "being the principal and almost only part so infested in all birds;" but why, then, are not all birds furnished with a similar comb?

There is yet a third theory—viz., that the said comb is used to straighten the vibrissæ with which the bill is furnished, and which may get clogged or bent in use. The singular tumbling over during flight, which White and Bishop Stanley think is due to the bird losing its balance whilst putting an insect into its mouth, might with equal reason be attributed to the action of combing out the vibrissæ. Here again a difficulty meets us; one of the American species, and the whole of the Australian genus *Eurostopodus*, have the rictus without bristles, and yet have the claw strongly pectinated.

All these theories are plausible, and I believe all may be occasionally exemplified. The feathers of the head may be infested with vermin, and the bird then naturally uses its foot as the only instrument whereby it can free itself from its tormentors, and so brings its toothed claw into use. Again, chafers and other large insects may be at times caught with the foot, in the manner of the kestrel; the long bristles of the rictus may become clogged or displaced in use; the foot again is the only means at the bird's disposal for straightening them, and in this the toothed claw may assist. I cannot think, however, that any of these uses is the one for which most of the species of this family are furnished by their Creator's wisdom with an instrument so peculiar.

My own observations have long led me to suspect another use. I am not sure whether the same idea has not been mentioned by some naturalist, though I know not by whom; but, apart from this, I am more and more convinced of its probability, and it has been still further confirmed by a minute examination of the foot of our European species, as well as some exotic ones.

The larger number of species composing the family of the Caprimulgidæ do not perch *across* a bough, as all other perching birds do, but *lengthwise;* and it is for this peculiar use that I believe the foot to be specially formed.

In the ordinary position of perching birds, the twig or bough is grasped by the foot, and thus a firm hold is obtained, the weight of the body, by tightening the tendons, increasing the stability; but it is evident the nightjar cannot effect any grasp of the bough as it sits lengthwise, and therefore the necessary firm position is obtained in other ways. It will be seen by any one who will take the trouble to examine it, that the serratures on the centre claws are therefore placed in exactly the best position for preventing the foot from slipping sideways; and this is still further provided against by the hind toe projecting forwards and inwards—so much so, indeed, that it has the appearance of being inserted on the inner side of the foot; and the whole organ is thus admirably adapted for its designed use.

It is worthy of remark, as tending to corroborate my theory, that there are one or two Australian species (*Podargus* and *Ægotheles*), and at least one in South America, which have the middle claw smooth and the hind toe directed backwards. This variation in structure leads, as might be expected, to a corresponding difference in the use of the organ; and we consequently find that they do not perch sideways, but across, and also hop from bough to bough.

I do not put this view forward as the undoubted solution of the disputed point, but think that from both positive and negative evidence the probabilities are in its favour.

There is a peculiarity about the jarring note of the nightjar which I have never seen mentioned by any writer. This peculiar sound consists of two notes, uttered alternately—one a third lower than the other—the highest being evidently made when the breath is expelled, and the lowest when it is inspired; and thus the jar is continued without intermission for a much longer time than would otherwise be possible.

Mr. Jesse, in his Gleanings, says, " they continue their

jarring note for a long space together, without seeming to draw breath," considering that the two notes are sounded during one expiration; but I have frequently heard their "jar" sustained for two, three, and sometimes four minutes, and I cannot conceive it could do this without taking breath; yet it is easily accounted for if my conjecture is correct. At any rate, the alternation of the two notes is so constant that it seems strange it has never been noticed.

I believe this jarring note is invariably uttered when the bird is perching, and never when it is on the wing, and it is not commenced for a week or ten days after its arrival. It has another short, sharp note, like the syllable "dek," which it utters during flight, and especially when any one approaches its haunts.

The eggs are usually marbled with light brown and ash colour on a white ground; but in 1856 I took two from a mossy hollow on the forest, in which the ground colour was yellowish white, marked with distinct spots of ash colour and brown without any streaks or marblings, the spots being accumulated at one end, and forming an irregular zone. At the time I found them the female was on the eggs, and sat immovably until I was within two yards of her, when she flew off, feigning lameness at first, and afterwards flew around me for some minutes as if desirous of intimidating me.

Of the family of the Columbidæ, three species are all that I can number—viz., the Ring Dove (*C. palumbus*), the Stock Dove (*C. œnas*), and the Turtle Dove (*C. turtur*).

Amongst our numerous woods and plantations the ring dove is plentifully distributed. It is an indiscriminate feeder on seeds of every kind; and, though well provided with abundance of beech-mast and acorns, it levies heavy contributions on the crops of the farmer. In one way or other, however, it has with us been pretty well kept in check, but the indiscriminate destruction of birds of prey is leading to a gradual increase in numbers, though they are not seen in those large flocks which in Scotland and the north of England have committed such serious depredations as to awaken public attention to the matter.

Few of our native birds are shyer in their habits, or more difficult to approach within gunshot. I have often walked to a clump of trees where a pair have been roosting, but they would invariably take flight from the highest part of the trees and as far out of danger as possible, never giving me the possibility of even a long shot, and this I have found to be their general habit. Mr. St. John, in his last interesting work,* remarks on the somewhat unusual tameness of these wild and wary birds, that they built in some shrubs close to his house and not above six feet from the ground, where, when sitting, they allowed the members of his family to pass without showing the least alarm. I have recently been told by a friend of a similar instance near his own house.

It would seem almost impossible for any bird to build a frailer nest than the wood pigeon, and I have often wondered that the eggs do not fall, or are not blown off by the wind from the slight platform on which they are laid, and through which you may sometimes see them from below. I remember one instance in which a pair had selected a young birch tree as the site for their nursery, the stoutest bough of which was not more than an

---

* Natural History and Sport in Moray. 1863.

inch and a half in diameter; but it was on one very much smaller than this, and close to the stem, that this apology for a nest was placed. It was formed of very slender birch twigs; but they had been so sparingly used that it was simply a piece of lattice work, through which the two eggs were distinctly visible from the ground, the distance being about twelve or fourteen feet. One of these eggs was very much larger than the other—a peculiarity I have noticed on several other occasions. I knew an instance in which a wood pigeon chose a very singular site for her nest; this was none other than a rabbit hole. A pigeon had frequently been seen coming out of this hole; the mouth was therefore stopped, and a cut made at some distance from it; on reaching the hole the hen bird was taken, with two eggs which she had laid.

In April, 1861, Mr. Sterling Howard communicated to the Sheffield Literary and Philosophical Society, a most remarkable instance of the selection of unusual materials for their nests by a number of domestic pigeons, and these were no other than horsenails! He said, "Over one end of the blacksmith's shop is a rude loft, in which are a number of boxes, the domiciles of the pigeons. The nails, which were abstracted from canvas bags and other receptacles, were of the ordinary kind for horseshoes, of various sizes, some new, others old and crooked. They were, however, laid with some regard to comfort, inasmuch as the points were not allowed to project upwards, but without any admixture of softer materials This is the more singular as there was abundance of straw, shavings, &c., in the neighbourhood. On these 'iron beds' the birds had laid their eggs, which were just ready for hatching when the discovery was made of the use to which the nails were applied. The nails, when removed, filled a watering-can holding about two gallons, one of the nests containing more than fourteen pounds weight."

The Stock Dove (*C. œnas*) is not uncommon, mingling, in the winter more especially, with the small flocks of ring doves. I never found the nest on the ground, where it is stated to be frequently placed; but I knew a very large and almost globular mass of ivy in the fork of an ash tree, in Blythe-corner Wood, in the close recesses of which a pair of stock doves reared their young for several years together. I do not think this species is so numerous with us as the preceding one, but it is still plentiful.

The latter word cannot, however, be applied to the Turtle Dove (*C. turtur*), one or two specimens being all that I have noted, though some have been killed in other parts of the county. I have never known it nest with us, and it can only be considered a rare straggler.

CHAPTER V.

GAME BIRDS.

THE Pheasant (*Phasianus colchicus*) abounds on all the estates in the forest district, and to such an extent that few would credit the immense numbers. They are almost as tame as barn-door fowls, and may be seen on the skirts of various plantations. Carefully tended and fed, and all their natural enemies destroyed, they become so accustomed to the presence of man that in many parts they will hardly take the trouble to get out of the way, and are scarcely entitled to the appellation of wild. Under circumstances so favourable they multiply rapidly, but a natural limit seems to be set to their increase, and frequently where they are most abundant large numbers are found dead without apparent cause. These are always exceedingly fat, and their plumage in the glossiest condition; they seem to drop down and die without a struggle. I have had them brought to me in this state, and have found their flesh plump and of good colour, and every feather smooth and perfect. The mortality from this cause is sometimes great, but it is only what might be expected when natural laws are interfered with; a farmer might as well attempt to keep an unlimited stock of sheep on his pastures. A remarkable instance of a similar result occurred in 1859 on Ailsa Craig. For some years previous to this date, every enemy of the sea-birds, ravens, falcons, &c., had been killed off by the tenant of the rock, that his own gains might be multiplied. Without their natural checks the sea-fowl increased beyond the resources which either the rock or the sea in the immediate neighbourhood could yield, and at the time of their annual migration in September, old and young died off in thousands, literally covering the sea for miles with their dead bodies. The grouse disease may be attributed to the same destruction of nature's police, the birds of prey, who if they had been left alone would have wed out the weakly birds, leaving the strongest to continue the race, whereas the sportsman does the contrary, killing the strong flying bird that rises first. When will game-preservers learn to value their best friends?

In addition to the ordinary plumage we have two varieties. All male birds possess a small tuft of feathers springing from over each ear, though a casual observer might fail to detect them; but a variety occurs in which these tufts (each composed of seven or eight feathers), are much lengthened, and form two prominent horns, especially conspicuous when the bird is excited, and giving it quite an imposing appearance. The most distinct variety is one with a ring of white round the neck, clear and well defined. It was introduced on the Rufford estate some years ago, and has so greatly increased that it is now commonly met with. Individuals irregularly pied with white are occasionally seen.

A very singular variety was shot several years ago near Edwinstowe, and was shown to me by the person who shot it; it was a cock bird, but the whole of the plumage was stone-coloured, with the usual black markings on the feathers.

I have been favoured by the Rev. C. Thompson, rector of Kirton near Ollerton, with some particulars of two birds which he shot in Kirton Wood, in October, 1865. From the description he has sent me, I think it most probable that they were hybrids between the blackcock and the pheasant. The first was shot in an isolated part of the wood, and proved to be a female. It was about the size of a blackcock, with naked legs and feet. The plumage on the back and neck was mottled brown and black, very similar to the back of the common snipe; breast and belly, brownish-white, not unlike the breast of a wild duck; tail square, with the two centre feathers rather longer than the rest, and the whole of them slightly tipped with white. Mr. Thompson had seen this bird in the same wood the previous winter, and she had also been noticed in the summer with two nearly full-grown young ones, in a cornfield near. She was killed on the 3rd of October, and on the following day Mr. Thompson shot in the same wood a male bird equally remarkable in its peculiarities. Its size was about that of a cock pheasant, but the whole of the plumage of the body was mottled black and brown; the head was black; neck, very dark glossy green; a ring of bright scarlet skin round the eye; iris black; tail similar in shape to that of a cock pheasant, but shorter, the colour dark brown, having each feather tipped with black; legs and feet naked.

Mr. Thompson thinks, and with great probability, that this bird was the offspring of the hen he shot the day before, one of the two young ones previously mentioned.

I much regret that these two birds were not preserved; but Mr. Thompson was not aware of their value as instances of hybridism.

UNIV. OF
CALIFORN

BLACK BANTAM HEN
in male plumage.

Naturalists have frequently noticed the fact that the hen pheasant will sometimes partially assume the plumage of the male. Montagu says, "in confinement;" and adds, "in a state of nature this circumstance probably does not take place. It is not, however, restricted to birds in confinement, for I have met with well-defined instances of it in wild birds. It is rarely that the full plumage of the male is assumed; but the change is most frequently confined to the head, neck, and breast, the black margins on the feathers of the latter not being so well defined as in the male, and consequently the contrast does not appear so rich. Sometimes the spurs are put on also.

In March, 1854, I met with a well-marked instance of this singular change. The head and neck were of the usual purplish green, and the bare skin around the eye bright scarlet; the breast and shoulders were glossy golden red, especially the former, each feather possessing the ordinary broad margin of velvety black. The rest of the plumage was entirely that of the female, but the change was further indicated by a small pair of spurs. This assumption by the female of the plumage of the other sex is well known to be produced by diseased ovaries, and is not necessarily connected with age.

I have before me the notes of a remarkable instance of transformation from the same cause in a domestic fowl of the black bantam kind, which belonged to my friend Mr. T. Sissons of Hull, by whom she was bred. Up to the age of seven years she laid abundance of eggs at all seasons, but never sat, or showed any wish to incubate. In her eighth year she ceased to lay, and began to assume the male dress, which in the following year was so fully attained, that no one ignorant of the change which had taken place would have had the least conception of her sex. Every characteristic of the male was put on; the comb and wattles became prominent, the long arched feathers of the tail, as well as the drooping hackles of the scapularies, were completely developed; and, to crown the metamorphosis, she had a formidable pair of spurs, an inch and a half long, and the scales down the front of the tarsi were very large. In this dress she lived until she was sixteen years old, when she died of a decay of nature.

Of the genus *Tetrao* I can include two species—viz., *Tetrix* and *Scoticus*. Black grouse (*T. Tetrix*) occur in scattered pairs all over the heathy parts of the forest, but in two localities (Inkersal Forest and Coleorton Corner) are rather plentiful. They are, of course, carefully preserved, and on the first-named spot as good a bag has been occasionally made as on some of the northern moors. Though fitted for inhabiting the ground, the blackcock is by no means a bad percher, and in several instances during spring time, when I have disturbed them during my walks, I have seen them fly off, making a circuitous route, and then settle on a bough of one of the old oaks, where they seemed quite at home. I have the eggs taken from the forest. My claim to include the Red Grouse (*T. scoticus*) in my list rests on a single male bird which was shot at Bevercotes in November, 1860, a solitary straggler from its northern home.

In the cultivated parts of the district the Partridge (*Perdix cinerea*) is most abundant, for it is closely and carefully preserved. It is, however, by no means confined to the arable lands, but is plentiful on the heathy parts of the forest, where the numerous ant hills furnish a rich supply of food for the young. Its affection for its eggs and young, and its devices to conceal them, or to entice intruders from the vicinity, are well known to every country resident, and often as I have witnessed its solicitude, it has always been with renewed interest. I once came suddenly on a brood of young ones, who could not have been more than a day or two old; they were accompanied by both old ones, and were busily feeding on an ant-hill in the midst of the moss and heather. On my unexpected appearance, the cock bird tumbled off on one side and the hen on the other, with well-feigned lameness. Out of curiosity, I threw myself on the ground and tried to secure some of the young ones; but, to my surprise, it was in vain. A few seconds before, there were ten or a dozen of them in a spot scarcely larger than my hat, but before I was down on my knees, they were dispersed in all directions amongst the surrounding heather, and I failed to capture one of them. I could not help admiring the instinct which prompted these tiny things to such instant and energetic action, for it could not have been acquired by imitation or experience.

The partridge is not readily disturbed from her eggs, but will sit closely until the last moment. One which had laid her eggs in the bottom of the hedge dividing my garden from the field, covered up her eggs every time she left the nest. This, I think, is not a usual practice; but as the garden was constantly frequented, it was no doubt adopted as a special contrivance against detection.

How singular it is that some of our native birds have become so accustomed to the strange sights and sounds of our railways; and how soon they have acquired the knowledge that to them they threaten no danger! I have seen partridges feeding in a field within ten yards of a railway, and taking not the slightest notice of a rapidly-passing train; and a little further on a common peewit was equally unconcerned. I once counted sixteen martins sitting on the telegraph wires at the Hitchin station of the Great Northern Railway; they not only manifested no alarm, but seemed actually to derive pleasure from the trains passing to and fro. Even the whistle of the engine failed to startle them. I have also known several instances of the house-sparrow building its nest in the ornamental iron brackets supporting the projecting roofs of some of the stations on the Liverpool and Manchester railway, where the traffic is almost constant, and where every train passes within four or five feet of the brackets. With the pert and familiar sparrow this is not so surprising, but with the others I have named, who are naturally timid, it offers a point of interesting consideration.

The Red-legged Partridge (*P. rubra*) has occurred in one or two instances with us, but they were mere stragglers.

The Quail (*P. coturnix*) is occasionally met with, but it is seldom that more than one bird occurs at a time. A friend of mine shot two in one day in 1848, in a field on the banks of the Maun, about seven miles from Ollerton. I cannot call this species a constant resident.

CHAPTER VI.

WADING BIRDS.

ONE of the regrets necessarily attendant upon progress and improvement, is that arising from the gradual extinction of much that once excited our admiration or interest. This is especially the case in the natural history of our country, where the advance in agricultural improvement, by draining our marshes and reclaiming our moors and forests, has been the means of banishing many races of animals which once were plentiful. Hence, while rejoicing at the increase in the material wealth of the country, and the general well-being of our population, the lover of nature cannot but regret the loss of some of our finest indigenous species of animals. The noble bustard and the stately crane have become matters of history as far as England is concerned, while many others are only known by the occurrence, at long intervals, of solitary individuals, who are no sooner seen than they are shot; and others again, though permanently residing with us, are by the gradual invasion of their haunts slowly but surely diminishing.

But though the great bustard may be considered extinct as far as England is concerned, its representative in miniature, the Little Bustard (*Otis tetrax*) occasionally appears, though only as a winter visitor. As such I am able to include it in my list, a single individual having been killed at South Clifton on the 21st December, 1866. It is singular that birds of this species should occasionally wander so far from their haunts on the plains of Southern Russia, and at such a season. Rarely more than a solitary bird is seen at a time, and it receives but a poor welcome.

Amongst the birds becoming scarcer every year is the Thick-knee (*Œdicnemus crepitans*), or, as it is also called, the great plover or stone curlew. Though scarce enough to be an object of interest, it is yet by no means a rare species, and regularly frequents many parts of our bare sandy forest land which are suitable to its habits. It used to breed on a large rabbit warren at Oxton, but the greater part of this has now been inclosed, and the thick-knee has disappeared with the solitude. I have known it also occur on Walesby Breck, and on the sheepwalks in the neighbourhood of Inkersal it may frequently be seen and heard. I have noted their arrival as early as the 17th of March. Its habit of resting in the daytime and squatting under cover of stones or bushes renders it difficult to detect in places which it is known to visit. It feeds and migrates in the night; every summer I have heard its well-known loud shrill cry, as it flew over my head in the darkness, most frequently during the season of its arrival.

The Golden Plover (*Charadrius pluvialis*) occurs in varying numbers, chiefly in early spring and summer. On the banks of the Trent it assembles in large flocks in winter, but in our own immediate neighbourhood I never met with it at that season, with the exception of two that were killed on the farm of Leyfields on the

10th of January, 1856; these were male and female. A few pairs breed with us every year, frequenting for that purpose the flood meadows between Ollerton and Clipstone. On Carbreck farm I have sometimes seen them in company with the lapwing.

The Grey Plover (*Vanellus melanogaster*) is occasionally seen in winter, but not in great numbers. One was shot a few years since by Mr. H. Wells in summer. This was on Edwinstowe Forest, and is the only one I have known of at that season.

The Lapwing (*Vanellus cristatus*) is common all over the district, and is to a large extent migratory, arriving with us in March, and leaving again about November. The wide extent of Thoresby Park is a favourite resort, as well as the moor-like parts of the forest and the large exposed fields on the forest farms. Here they regularly breed; and though the eggs are sometimes collected, they are not found in such numbers as to make it so profitable an employment as it is in some of the eastern counties. I have often been misled when in search of the eggs by the enticements and manœuvres of the parent birds, notwithstanding my knowledge of their practice. While the female is on the nest the male is seldom far distant, and on the approach of an intruder wheels and tumbles around very closely, in order to entice him from his partner. The latter on leaving the nest runs quickly for a short distance before taking wing, and by this device renders it difficult to detect the exact spot she has left, and this is increased by the colour of the eggs themselves.

Few birds are more in keeping with a retired country scene, or add more to its picturesque beauty, than the Heron (*Ardea cinerea*), whether it is a winding sedgy stream, or the wider expanse of lake or mere in which it takes its stand. The attitude in which it is generally seen is one of pensive quietude, for unless the observer is an early riser, or a watcher in the dim evening twilight, he does not behold it in its more active moods.

The island on the large sheet of water in Thoresby Park was long the resort of four or five pair of herons, who built their nests on the tall trees with which it is thickly covered. Here in the daytime some were generally to be seen in watchful inactivity, sometimes standing in the shallow water a few yards from the bank, but more generally roosting on the trunk of a large silver willow, which, growing on the margin of the island, had given way and fallen until it lay at a slight inclination, or on the large projecting arm of another that grew close by. Though the island is a long way from the shore, and very far beyond the reach of a gun, they invariably took flight if any one stood on the mainland opposite; and it was a pretty sight to watch them wheeling high in the air, or flying off to a wood on the further shore of the lake, where on the tops of the highest trees they would perch to keep guard against their enemy.

I have often approached to the edge of the shore under cover of the thick shrubs, and with a telescope have been delighted to watch their movements. It has surprised me to see that even those that appeared the most listless and unconcerned were extremely vigilant, their bright eye marking everything that moved around. Sometimes a mallard or a teal would come flying by, with a loud warning "quack," when the herons would be on the *qui vive* in an instant, ready to take wing at the smallest sign that indicated danger.

It is amusing to see them perch on the top of a tree;

their long legs appear as if quite in the way, and they sway backwards and forwards like a pendulum for a few times until they have found the centre of gravity.

This little colony never increased in numbers; for some years no young were reared at all, the eggs being destroyed by the carrion crows, who seemed to have a particular spite against them. I have seen a heron when flying over the island, give active chase to a hooded crow that came by, pursuing it fiercely, and uttering shrill cries, as if aware of hoodie's propensity. They cannot, however, suffer from its depredations, as the hooded crow leaves us for the sea-side before the eggs of the heron are laid.

The herons on the island did not by any means confine themselves to Thoresby, but frequented the lakes in Welbeck and Rufford Parks, and the surrounding streams. I have written of them in the past, for I regret to say all the birds were shot by the keepers in 1856. Their depredations were so great, both in the lake and in the streams, that it was determined to sacrifice them, and one by one they became victims to the proscription.

A few years since a heron was found dead on the edge of a fish-pond at Walling Wells, the seat of Sir T. W. White. The way in which it lost its life was very singular. In the pursuit of its prey it had struck its bill completely through the body of a large eel, near the head, without immediately killing it, but the eel in its death struggles had coiled itself round the neck of the heron, as well as round some aquatic plants on the bank, and both were found dead in that position. Curiously enough, an exactly similar occurrence was recorded and figured in the *Illustrated London News* in January last.

My notices of the next six or seven birds are very meagre, for they are mostly stragglers in our district, and some indeed but occasional visitors to our shores.

I have notes of the Bittern (*Ardea stellaris*) occurring on four occasions; one in 1846, shot in a willow holt on the banks of the Trent, near Nottingham, a second in the next garden but one to my own in 1853, a third on the water at Carburton in 1863, and a fourth in a boggy place on the margin of a small stream in Welbeck Park in 1866. We have few, if any, haunts suitable for the permanent residence of this handsome bird, and indeed, throughout the country there are comparatively few spots where now can be heard what Scott graphically calls

"The bittern's sounding drum,
Booming from the sedgy shallow,"

and they are yearly becoming scarcer. The bird I have mentioned as killed near my garden was shot in mid-winter, during a long and hard frost, and was little more than skin and bone. It had fared badly indeed, and had lost its fear of man, for it made no attempt to escape when perceived by the person who shot it..

A few specimens of the Curlew (*Numenius arquata*) are occasionally seen on Inkersal Forest during the time of their vernal migration, but I am not aware of their breeding there; I have also seen a solitary one on the edge of the lake at Thoresby. Every spring small parties may be seen in the daytime passing high over head on their way to their breeding places on the Yorkshire moors; oftener at night I have heard their well-known clear shrill whistle, uttered by one of a party when it has been too dark to see them, and in a few seconds the cry was taken up by one and another as they passed, and had a peculiarly wild effect.

The Redshank (*Totanus calidris*) is the next of our visitors, but at rare intervals. I have only known of three, all occurring in the winter. One was killed in a small boggy piece of ground bordering the stream on the outskirts of the village; another, a young female, was shot in 1859, in a meadow on Carbrecks farm, while feeding.

The Common Sandpiper (*T. hypoleucos*) occurs sparingly on our streams, but I never met with more than a pair at a time. A low-lying meadow a short distance from my garden was frequently chosen by a pair as their summer residence, and their lively habits added a charm to the quiet stream on whose banks they fed. If left unmolested they would doubtless have bred there, but they generally, alas! fell victims to the mania for shooting everything strange.

My claim to include the Greenshank (*T. glottis*) in my list rests on the occurrence of a single individual seen by my father many years ago on the same piece of swampy ground which I mentioned as having sheltered the redshank. It rose on his approach, and flew slowly away, and was not seen again.

The same remark applies to a bird which is still more seldom seen—the Avocet (*Recurvirostra avocetta*). On the 24th of July, 1856, one was seen in a meadow on the banks of the stream at Edwinstowe by a boy, who of course did not know what the strange bird was. He managed in some way to steal upon it so closely as to kill it by a stroke of a stick. It proved to be a young bird, in good condition, but in immature plumage. It is not often that this bird comes inland; even when more plentiful than it now is, it was chiefly on the low mud flats and marshes bordering the sea that it was to be found during its summer visits, and on the latter that it deposited its eggs. The bird I have mentioned may have been bred in the Lincolnshire marshes.

Of still greater rarity is the Black-winged Stilt (*Himantopus melanopterus*), and I am pleased to be able to note its appearance, though only on one occasion. This was on the 30th of January, 1848, when one was seen standing in a ditch of shallow water in an ash-holt at Perlethorpe, by Mr. Mansell and his father. It was apparently feeding when they caught sight of it, and on being disturbed rose with a shrill "squeal," flying low in the direction of the river that bounds the ash holt. Mr. M. would have shot it had not his father stood in the way, so it continued its course unharmed. Its extraordinary long legs gave it a most singular appearance when standing in the water, and still more so during flight, for they were then carried stretched out behind, and the clear contrast afforded by its black and white plumage and red legs was very striking. It was doubtless a visitor from the Continent, for this species has, I believe, been seldom seen of late years in this country.

The Black-tailed Godwit (*Limosa melanura*) and the Bar-tailed Godwit (*L. rufa*) are occasionally met with; more frequently the former, which is not uncommon on the banks of the Trent in the neighbourhood of Nottingham; but both are but stragglers.

The Woodcock (*Scolopax rusticola*) is generally considered as a winter visitor, retiring in the spring to Northern Europe for the purpose of rearing its young, though many occasional instances of its breeding in various parts of the country have been noted, especially of late years, more attention having been paid to the subject. In our own forest district, however, they have become constant denizens throughout the year, breeding regularly in our woods in great numbers. I have no doubt that these are regularly augmented and diminished by a partial migration from and to the Continent in October and March, but still a very large number take up their permanent residence with us, and may be constantly met with during the summer. Ollerton Corner, and Blyth Corner Woods abound with them to such an extent that, while walking along the side of the former wood on a summer's evening for the distance of a mile, I have counted at least from twenty to thirty woodcocks flying from the wood to the forest, and this I could do any evening during twilight. In these woods they breed abundantly, their nests being loosely formed of dry leaves and fern, with sometimes a little grass, and placed in a warm sheltered situation. The eggs are always four in number, the young being hatched about the last week in April or the first in May, and being able to run as soon as they leave the egg.

The female shows the greatest affection for her progeny when very young, hissing in a menacing manner on the approach of an enemy, and when compelled to retreat taking one under her wing, and sometimes one under each wing, and conveying them away to a place of safety.

In 1846 a woodman was engaged with some others in clearing the underwood from a plantation in Thoresby Park. He had just cut down a small thornbush with his billhook, when directly it fell, a woodcock started up from the fern at its foot, where she had been brooding four young ones. She fluttered round them for a few minutes in great alarm, and then taking up one under one of her wings she ran off with it for a short distance, beating the ground with the other wing in the manner of the partridge; the remaining three young ones quickly concealing themselves in the grass and fern.

A friend of mine, who was a very close observer, informed me that he came on a nest in a wood called the Catwins, where the young, four in number, and only recently hatched, were being brooded by the female. On being discovered she did not stir, but hissed at him like a goose, in order to deter his approach. His curiosity prompted him to advance within a yard to see what she would do, when she merely moved off about the same distance, followed by her young family, and again brooded them. The tiny things were covered with blackish down, very similar to the young of the waterhen, and with their long bills looked very funny little fellows.

A week before this Mr. Mansell found a nest with four eggs in a dry place at the foot of a furze bush in Thoresby Park; but, contrary to the usual habit, it was in a most exposed situation, being only about five yards from the public carriage-way leading to Clumber, and no attempt at concealment had been made. These eggs were safely hatched a few days afterwards. A nest which I found in Blyth Corner Wood consisted merely of a layer of dead leaves, with a few pieces of dry fern, and was warmly placed in a dry spot at the foot of a bush amongst various trees of large growth, and within a short distance of a small shallow pool.

These instances of the woodcock's breeding were, of course, only accidentally discovered in closely-kept preserves, any curious and inquisitive deviations from the rides and paths through them which a naturalist might be inclined to make being forbidden, except you are one of the privileged ones. Nevertheless, the frequency of their breeding with us can be fully proved any summer's evening by the numbers which may be seen as they leave the woods for their feeding grounds.

An interesting fact in the natural history of the woodcock has lately been cleared from a doubt which hung around it; I refer to the question whether it utters an alarm note, or whistle, when flushed in covert. For myself I had no more doubt of this fact, than that it utters an ordinary call in feeding time, or during the season of pairing, for I have repeatedly heard this note of alarm when aroused. Many appear to have thought that the bird rises mute, and strange to say many sportsmen were of this opinion, but others of great experience, and who have combined the observant character of the naturalist with the energy of the sportsman, have clearly testified to the fact that an alarm note is uttered by the bird on rising.

In the beginning of January, 1859, a woodcock whose plumage was entirely white was shot in Thoresby Park.

The Common Snipe (*S. gallinago*) is numerous wherever the ground is suitable to its habits. It is chiefly a winter visitor, though birds have been killed in summer, and I have been told on good authority that its eggs have been found on one or two occasions, but I have not seen them. In severe winters they are very fearless, and I have noticed them close to my own garden. During a hard frost, when the ground was covered with snow, I started one from a drain in a low-lying meadow, three days together, and within a few yards of the same spot, the bird showing little signs of shyness, and allowing me to approach closely before taking wing.

The little Jack Snipe (*S. gallinula*) is occasionally met with in the winter. Like the larger species, it will frequently return to the spot from which it is roused. It seems possessed of a secret worth knowing—viz., how to live well, for it is always in good condition, and never seems to suffer even in the hardest frosts. In the severe winter of 1849-50, when fieldfares and redwings, through the long continuance of the frost, were so greatly emaciated as to be little else than skin and bone, allowing themselves to be approached within two or three yards, a jack snipe was brought to me which surpassed any I ever saw. I skinned it for preserving, and found its whole body covered with a layer of solid fat to the depth of a quarter of an inch—not a bad protection against intense cold.

That bird of singular habits and note, the Landrail (*Crex pratensis*), visits us in abundance every year, sometimes arriving as early as the 1st of May, while in 1853 I did not hear its note until the 18th. This was unusually late, the season being a remarkably cold and backward one, a fact of which our other migratory birds also seemed, in some mysterious way, to be fully cognizant. Nothing, indeed, relating to the feathered tribes is more wonderful or more deserving of our admiration than that knowledge, call it instinct or what you will, which, implanted in them by their Creator, enables them to hasten or delay their departure for their distant but temporary places of abode, according as the seasons there are suitable to their necessities or otherwise. How strikingly is this wisdom brought forward in Holy Scripture to shame man's neglect and ingratitude!

"Yea, the stork in the heaven knoweth her appointed times; and the turtle, and the crane, and the swallow observe the time of their coming; but my people know not the judgment of the Lord."

I have never succeeded in causing the landrail to take wing except with a dog, and even then its flight is always brief, as it takes an early opportunity of dropping to the ground and regaining its covert. It flies rather slowly, with its legs hanging down, and there is such an air of effort about its movements on the wing, that I have wondered how its migrations are performed.

On the ground, however, all this is reversed. It is marvellous to see the ease and rapidity with which it threads its way through the corn or grass; and even when the latter is short, as it sometimes is on the landrail's first arrival, the bird's course is so smooth and well concealed that only now and then you perceive any motion of the grass to indicate its whereabouts.

Its form and plumage are admirably adapted to its habits and requirements; none of its feathers project beyond the graceful outline of its body, and they are particularly close and firm in texture. When in motion the head and neck are carried in a horizontal line with the body, the whole constituting a most efficient wedge, enabling the bird to thread its way through the densest foliage with the greatest facility.

Its ventriloquial powers are well known to every observer. Now its harsh "crake, crake," seems within a few yards, and the next moment it sounds as if it were half across the field, and this apparent variation in distance is so well simulated, that in a consecutive repetition of its call for ten or twelve times, a few notes will sound as if uttered almost at your feet, and the next two or three from afar, and yet the bird is standing motionless all the time, as I have several times tested. Its singular call I have often imitated by drawing my nail across the teeth of a pocket-comb, and thus inducing its near approach.

The female sits very closely on her eggs—so closely, indeed, as not unfrequently to lose her life by the mower's scythe ; I have known two instances of this, in one of which the poor bird was almost cut in two. A single specimen of the Spotted Crake (*Crex porzana*) was killed in a swampy piece of ground by the stream at Budby, in October, 1863. I have known no other instance of its presence with us.

All the streams which intersect our neighbourhood, and especially the large lakes in the parks, are abundantly stocked with the Water-hen (*Gallinula chloropus*). In the streams it only frequents the stiller reaches, where the banks are fringed with reeds or bushy aquatic plants, which afford it concealment when required. It is, however, by no means a shy bird, and where its haunts are in the neighbourhood of houses it becomes comparatively bold. At Budby the stream washes the side of the turnpike road, which on the other side is bounded by a row of houses ; here the water-hens mingle with the ducks and geese which throng the water from a neighbouring farmyard, and, regardless of passers-by, walk fearlessly on the road, feeding amongst their tame companions. In Thoresby Park they are equally tame, frequenting the lawns and gardens, and the park itself in the vicinity of the water, especially near the bridge leading to the house, and running about amongst the sheep and deer. A few pairs inhabit a deep still part of the stream flowing through our village, the banks on either side being occupied by gardens. Here they are fond of making incursions, and in the summer time of feeding on the currants and gooseberries; but they run to the water on the slightest alarm, and, if unable to conceal themselves amongst the weeds on the water's edge, will sink their bodies until only the beak is above the surface, remaining quite still until the danger is removed, and rarely making any attempt to escape by diving. As my own garden abuts on this part of the river, I have often carefully watched them when they have taken to the water as I have described, and have twice detected them in this trick by the bill when the body was submerged, but it required a very close scrutiny. On one of these occasions the bill was projected in the midst of a few broad flaggy leaves of grass, which were bent down upon the water.

The Water Rail (*Rallus aquaticus*) is not uncommon, but only on some portions of our streams which are peculiarly suitable to its retiring habits; these are chiefly swampy ash-holts at Perlethorpe, and a similar marshy spot much covered with alders at Budby; here, however, they must be sought for. They are rarely seen swimming on the streams themselves, but delight in wading in the shallow pools and amongst the dense aquatic vegetation.

Spots such as these do not tempt the ordinary excursionist, and few except the sportsman and the naturalist care to risk a wet foot in their investigation. The water rail is most plentiful in the first-named locality, which is especially retired, and is the same where the black-winged stilt was seen which I have previously mentioned. I have often searched for, but never succeeded in finding, its eggs.

In March, 1849, a friend of mine picked up a water

rail, which had killed itself by flying against the telegraph wires on the Hull and Selby Railway, on the banks of the Humber.

The Coot (*Fulica atra*) is a plentiful species on all our sheets of water, but is especially abundant on the large lake which is so great an object of beauty in Thoresby Park, and which contains nearly one hundred acres. It is about a mile in length, and on the southern side is bounded for about two-thirds its extent by a wood called the Lawn Plantation. A strip of this along the bank, covered with grass and reeds, and planted with shrubs, is allowed to grow undisturbed, and is a favourite resort of the coot at all times of the year. Here they breed without molestation, as well as on the island which is just opposite, and which I have mentioned as the breeding-place of the heron.

The nest is usually formed of a large mass of rushes and flags placed on the water amongst the growing reeds, but raised a sufficient height to keep the eggs dry; it is sometimes, though more rarely, placed on the ground, and I once found a nest, containing nine eggs, which was built on the trunk of a tree which grew on the bank of the island, but which had gradually fallen over until the trunk rested on the surface of the water. On this trunk, between two small boughs projecting upwards, a coot had piled a large quantity of flags. This was on the 13th of May; and while rowing down the side of the island we found, behind a projecting point, seven young ones, which could not have been hatched above a couple of days; they were accompanied by the mother, but we came on them so suddenly that she, with great alarm, took to instant flight, leaving her young family. With these we were greatly amused; for though they kept cheeping piteously for their absent parent, they showed not the slightest alarm at us, but as we rowed quietly along they swam slowly after our boat, and so closely that I took up one in my hand, and as we left the island we had to drive them away, or they would have followed us to the shore, which was at some distance, and where, possibly, the old bird might not easily have found them. The little things, with their dingy bodies and reddish heads, looked so pretty, and showed such confidence in us, that I felt quite sorry to leave them. I have known the male take charge of the young, tending them as carefully as the female; whether this was owing to the death of the latter I cannot say, but I have noted an instance of the male being shot when thus engaged. The eggs of the coot are very constant in colour and markings, a cold stone-coloured ground uniformly covered with small blackish dots and specks.

I have never but once seen a coot on the land; and here, as might be expected, from the construction of its feet, its actions were rather ungainly, though not so much so as might be supposed; on the water it is quite another creature, thoroughly at home, diving incessantly, and sometimes to the amazing distance of at least ninety or one hundred yards. Its skill as a diver is best seen in pairing time, when the male amorously chases the female, sometimes on the surface of the water, but as often beneath, and under cover I have watched with much interest and surprise the extent of their sub-aqueous evolutions, which I could see distinctly.

CHAPTER VII.

WATER BIRDS.

AT first thought it might perhaps be concluded that in a strictly inland district, comparatively few species of birds of purely aquatic habits would be found. In this locality, however, there are features which tend materially to modify this conclusion. The numerous parks which have given the forest district the title of the "dukeries," are most of them graced by artificial lakes of greater or less extent. That at Thoresby covers about ninety-five acres, and those at Clumber and Welbeck are not much smaller; while Rufford and some others are more contracted. These lakes, with the streams supplying and flowing from them, lie in the greatest seclusion, and form secure and quiet sanctuaries, which offer such attractions that, in addition to their ordinary residents, few winters pass without the presence of some of the rarer species of water-fowl. The river Trent also forms a sort of highway from the sea, by which many littoral and pelagic species find their way into the district; and it is chiefly thus that we can account for the occurrence of the solan goose, the little auk, the terns, and other true sea-birds amongst those which visit us occasionally; and the goosander, the smew, and some of the gulls as constant winter residents.

The strictness with which the whole district is preserved by the respective owners, secures that seclusion which is so congenial to the habits of wildfowl. For many a mile their haunts are undisturbed save by the occasional visit of the keeper, or the foot of some prying naturalist; and many a quiet summer hour have I passed in observing the wild duck

> "Lead forth her fleet upon the lake,"

or in watching the gambols of the little grebe and the water rail, or the glancing flight of the kingfisher; while a walk in winter would reveal in addition the pochard and tufted duck, or a flock of the handsome but wary goosander.

There is one subject connected with this order that has often excited my interest and occupied my thoughts —viz., the property which their feathers possess of resisting wet. I have never felt satisfied as to the correctness of the commonly received theory that their repellant qualities were owing to a dressing of oil which the bird applied with its bill, and which it obtained from the gland or glands which are situated on the rump; the more I thought over it and the closer I observed and examined, the more convinced I became that the idea was not based on fact. That the feathers of such birds as seek their food on the water *are* waterproof needs no demonstration; and this is especially the case with those whose home is on the sea, and whose feathers are almost oily to the touch.

Some birds are furnished with only one oil gland on the rump, while others have two. I have taken some pains to discover how this distribution is made, and from an examination of a large number of species I have ascertained, as far as my inquiries have extended, that the possession of one gland is confined (with some exceptions, which I will mention presently) to those birds which are strictly land feeders; while those provided with two seek their food on the water.

It may, perhaps, be considered that this fact goes to strengthen the common opinion; but let us examine it a little closer. First, it is necessary for the maintenance of this theory that the gland or glands should contain a supply of oil sufficient for the constant requirements of the birds; and this, in those whose time is spent upon the water, must be very large, for a daily application at least would be needful to keep the feathers as repellant as is requisite, to enable them to obtain their food and to preserve that warmth and dryness which is essential to their existence. But let any person ascertain by actual experiment the quantity of oil contained in one or both glands, and it will be seen at once that it is totally insufficient for the alleged purpose, even in those water birds which have the glands very large, as the divers and grebes. Willughby, I think, says that these species, and such as want tails, have the glands small; but my experience is quite the contrary, and I believe that the tail has nothing at all to do with the question.

But there is another difficulty. Supposing the ordinary theory to be true, then all birds, and especially the aquatic species, would make constant application to the glands during the process of preening their feathers, in order to obtain thence the oil for dressing them. This I can venture to affirm is not the case. I have watched various birds, both wild and tame, during this operation, and certainly never succeeded in witnessing it. The feathers on the rump were dressed in the same way as those on the other parts of the body, but there was no repeated application to the gland, as must have been the case if it had been necessary to obtain thence a supply of oil. Let any one watch a duck thus engaged, and they will bear me out in this; and yet at the next plunge into the water the feathers are as oily and repellant as ever, and the drops of water shoot off them like molten silver; the feathers on the head, too, are as repellant as those on the body; and yet it is evident that the bird cannot possibly apply oil to that part.

It has often struck me as very strange that one writer after another should have gone on repeating the same story, without apparently taking the trouble to examine and test its truth. There may, indeed, be some little plausibility about it. All birds, from a natural love of cleanliness and personal comfort, preen their plumage; but water birds are more assiduous in this respect than land birds, and for a very good reason. It is absolutely necessary that their plumage shall lie very closely, both to diminish friction during their passage through the water, but more especially to prevent the latter having any access to their bodies, thus increasing their buoyancy and maintaining their warmth. To a water bird, therefore, a broken and disordered feather is of the utmost importance; it is essential that all shall be smooth and compact; and it is to effect this that they are so frequently seen trimming them and passing the webs through their bills; and this act has doubtless given rise to the supposition of a practice which I believe to have no existence, and which, with few exceptions, has been repeated by every author, from Montagu to the present time.*

The gallinaceous birds have generally only one gland, which, as they do not seek their food upon the water, would, according to the common theory, furnish sufficient oil for their wants. But surely a dressing of oil, however slight, would be a bad preparation for the practice of dusting, in which these birds love to indulge. Instead of tending to their cleanliness and comfort, it would have a contrary effect, dirtying and clogging their feathers. And yet we know it is not so; for, after the dust is expelled by a few vigorous shakes, their plumage is as clean as ever.

The practice of dusting, which seems merely to be employed as a simple mechanical means of dislodging the parasites with which all birds are infested, I have seen practised three or four times in succession; and if the feathers had been oiled, this repeated application of dust would but increase their filthiness.

While thus rejecting the common notion, I would not venture to assert that I am *certain* of the true use of

---

* "Nature, ever provident in all her ways, has taken care to supply every bird, more or less, with an external secretion of an unctuous nature, situated in a glandular bag upon the rump, which they instinctively make use of for oiling and dressing their feathers as occasion requires. In water-fowl this bag is most conspicuous, and it is remarkable that birds most frequently use it after washing, previously to their feathers becoming perfectly dry."—*Montagu's Ornithological Dictionary*, p. 136. "And finally, the gland which all birds have at the rump, and from which they express an oily matter to preserve their feathers moist, is most considerable in those that live upon the water, and contributes to make their plumage impermeable."—*The Sea and its Living Wonders*, by Dr. G. Hartwig, p. 123. Longmans, 1860.

the glands, but I have no hesitation in coming to a conclusion which, to my mind, is perfectly satisfactory, and that is, that they are simply excretory. In the course of my examination of various species, I found that not merely were the webs of the feathers of aquatic birds oily, but that the shafts and quills were equally so. The chief food of most pelagic species is of an oily nature; and their stomachs are filled with oil to such an extent that, as is well known, the body of a fulmar petrel, with a wick of cotton drawn through it, is the common lamp of the inhabitants of St. Kilda. I have also often seen the Cape pigeon (*Daption Capensis*) vomit nearly a tablespoonful of clear oil when captured. It is evident that in these and similar species which seek their food on the water, no outward application of oil to their feathers is necessary, for both skin, flesh, and feathers are thoroughly impregnated with it.

In support of my view we find, as I said before, that those birds whose food is obtained from the water, and whose flesh and skin are more or less oily, are furnished with two glands, in some species of large size; whilst land birds, whose flesh is generally free from oiliness, have only one gland; and in some of these, as in the black grouse for instance, it is very small, though singularly enough, the red grouse and ptarmigan have each two glands, but very small. An apparent exception exists in the white-tailed sea eagle, which has, like the rest of the rapacious birds, but one gland; but though it feeds on fish, which it captures in the water, lambs, hares, and other animals also enter largely into its diet. But there is another more remarkable exception to this arrangement, and one totally incompatible with the popular idea, and that is the group of the penguins.

P

These birds are more thoroughly aquatic than any others, seeking their food entirely under water; and, if the received theory is true, they would require much oil to keep their plumage waterproof, notwithstanding the scale-like character of some of the feathers; and yet, strange to say, they do not possess the vestige of an oil gland, and they, consequently, have no means of " oiling their plumage."

I had some correspondence with Mr. Zurhorst, the eminent poultry breeder of Dublin, on this subject, who considers from his observations on his domestic ducks, that they *do*, or at least appear to, apply the oil which they obtain from the gland, to their feathers. He says, " At all events the feathers around the glands are raised on end, the bill is buried in them, accompanied by a jerking motion expressive of pressure or squeezing being used on some special part or parts; this is followed by repeated applications of the bill to the different parts of the plumage—not at random, but going carefully over the surface, inch by inch. My Aylesbury ducks, after long confinement, when fattening for exhibition purposes, when turned out on the pond for the first time, are evidently loose-feathered; the plumage becomes speedily saturated, and, if the birds are at once confined again, takes very long to dry; whereas if, after being allowed time to dry and dress their feathers, they are turned in again, they come out apparently as sleek and impervious to wet as ever they were. I would incline to think that if the oil were supplied by an involuntary operation through the body itself, there would be no deficiency in the supply during confinement. With reference to gallinaceous birds, the dusting process strikes me more as a cleansing process than specially directed against parasites, and as being performed for the purpose of cleaning the feathers from any cloggy matter that may be on them, including the effects of natural perspiration; the process of pluming themselves being by no means so frequently repeated as in waterfowl."

I am afraid Mr. Zurhorst has mistaken, as I believe has often been done before, the preening which ducks give to their feathers for the alleged process of "oiling." I have made careful observations with an especial reference to this question, for the last fifteen years, and I never once succeeded in witnessing any duck make repeated application to the oil gland; in fact, it was from watching domestic ducks that I first became convinced of what I consider the fallacy of the common opinion. I have seen them trim the feathers around the glands with as much care as any others, but not more so; and, as Mr. Zurhorst says, I have marked them go over their feathers "inch by inch." This assiduous preening of the feathers by the domestic duck is greatly surpassed by wild birds, and both have good reason for the practice. It is absolutely essential, both for warmth and dryness, that their plumage shall lie close and compact— a disordered feather is therefore all-important, and, no matter how oily the plumage may be, such a feather would admit the water, and hence the necessity of careful preening. Nothing can surpass the trim neatness of wild ducks, for instance, and I have been surprised at the time spent by them, as well as by widgeon and teal, over their toilets.

Mr. Zurhorst's Aylesbury ducks are an illustration in my favour. All birds thus fed up in confinement are in an unnatural condition, and more or less unhealthy—

witness the enlarged livers of fattened geese. Deprived of access to water, on which nature intended them to live, they do not feel the *necessity* for a careful preening of their feathers, and, as a consequence, do not practise the habit as they did before; but let them be turned out to a pond, and the natural habit is resumed with the returning necessity. I have myself repeatedly remarked this loose plumage of ducks that have been put up to feed, and how little they preen their feathers.

I am glad to find my views confirmed by the late Mr. St. John, who, as a practical and accomplished naturalist, was well qualified to pronounce an opinion on such a question. He says:—

"The imperviousness to wet of the plumage of wild-fowl is evidently not caused by any power which the birds have of supplying grease or oil to their feathers. The feathers have a certain degree of oiliness no doubt, but from frequent observation I am convinced that it is the manner in which the feathers are placed which is the cause of the water running off them as it does.

"As long as a wild duck of any kind is alive, his skin remains perfectly dry, though in the water, and although, from the situation in which he may be placed—being pursued, for instance—it is quite impossible for him to find time to 'oil his plumage,' as some authors assert he does, 'in order to keep out the wet;' but the moment a duck or water-fowl is dead the water penetrates through the feathers, wetting the animal completely. If one wing is broken, the feathers of that wing immediately become soaked with wet, the bird not having the power of keeping the feathers of the broken part in proper position to resist the entry of the water. We all know that birds are able to elevate, depress, and, in fact, to move their feathers in any direction by a muscular contraction of the skin. When this power ceases, the feathers hang loosely in every direction, and the wet enters to the skin.

"The live otter's skin never appears to be wet, however long the animal may remain under the water, but, like the plumage of birds, soon becomes soaked through when the animal is dead. Whilst he is alive the water runs off his hair exactly as it does off the back of a bird during a shower. When we find any live water bird or animal with its feathers or hair wet or clinging together, it is a sure sign that the creature is either diseased, or is suffering from some wound or accident."*

Cage birds often suffer from ulceration of the rump gland, which affects the general health of the bird, as the stoppage of any other of the secretions would do. This doubtless arises from the unnatural condition in which they are kept, and not, as Bechstein amusingly says, from their neglecting to use the gland.

I will now enumerate the true water birds which have come under my observation. Of many of them my notes are very brief—a mere mention of their occurrence; but it must be remembered that it is not my object to write a general history of the respective birds, but to record observations actually made in a particular locality. My readers must therefore kindly make allowance for, in many instances, the paucity of information I have given.

Of the Anserinæ I can only include two species. The Grey Lag Goose (*Anser palustris*) is not of unfrequent

---

* Natural History and Sport in Moray, by the late C. St. John, p. 65.

occurrence. I have met with it both on Thoresby and Rufford Lakes, though it cannot be looked for regularly, as we have none of the marshes in which it delights, and where it can be secure from interruption, for it is excessively shy and wary, and detects a sportsman at a great distance. In winter I have often seen large flocks of this species passing overhead from the northward in the well-known form of the letter ◁; but it is only in twos and threes that they pay their brief visits to us.

Of the Bean Goose (*A. ferus*) I have known a few individuals obtained, but they do not visit us in any numbers as in the neighbouring county of Lincolnshire, and their well-known wariness makes it very difficult to approach them unobserved.

That fine winter visitor the wild swan or Hooper (*Cygnus ferus*) appears in small parties in hard seasons, generally frequenting the Trent, where, I am sorry to say, it soon falls a victim; for no sooner is a flock seen than numberless guns are put in requisition. I have known several instances of their occurrence on the river near Nottingham; on one occasion the flock consisted of seven, and all of them were shot. Two Hoopers visited Thoresby Lake in December, 1863. They did not seem to fraternize with the next species, and were both shot by the keepers.

The artificial lakes I have mentioned are all tenanted by the Mute Swan (*C. olor*). They are most numerous on the sheet of water at Thoresby, where I have sometimes counted more than thirty at a time. Their numbers vary a little from time to time, small parties of two and three arriving from Clumber or Welbeck, and their visits being returned in due course. I once saw the arrival, on Thoresby Lake, of a pair of the Australian black swans which are kept by the Duke of Portland at Welbeck; they flew very rapidly, dashing energetically into the water, and continuing to dive for a minute or two—a habit which I never saw practised by the mute swan.

A number of the latter species were on the lake at the time, but their old world notions seemed quite shocked by the vigorous gambols of the colonial birds. They raised their snowy pinions, and arched their necks with increased dignity, as if to remind their sable relatives that they had quite forgotten the proprieties of swan life; but they vouchsafed no further recognition of them, and gradually sailed away to another part of the lake.

The female makes use of the same spot for many years in succession, and I have known three such places at Thoresby which have been occupied without intermission for at least a dozen years. One of these is a small island in the river near the house, just above where it is crossed by a bridge; the island is covered with large trees, but as it is only a few yards from either bank, it is very much exposed to every passer-by. Here the nest is always composed of decayed sticks, which have fallen from the trees, and has little else intermingled with them. Two other spots in the pleasure grounds have been selected for nearly an equal length of time, and in both these sticks were used for the nest; in two other places it was constructed of flags and rushes—if we can say so of a mere heap of those materials. The female alone collects the materials for her nest, the male never condescending to help his partner, but contenting himself with keeping watch and ward, and is ready to do battle with all comers, great or small. It is not necessary to make any burglarious attempts to be entitled to be considered an enemy; the wayfarer may pass along near if he go on his way like a peaceable person; but if he stand to look, the jealous husband's ire is immediately aroused, his wings are raised, the feathers of his neck are erected until it becomes double the usual thickness, and instead of his ordinary quiet gliding motion, he propels himself by violent strokes at long intervals. The observer had better beat a retreat now, for if he delay another minute, the bird will fly out of the water at him, and a stroke from his wing is no joke. I confess that I have often ignominiously run away under such circumstances, when from motives of curiosity I have tried his patience too long; and I was even once attacked when on horseback.

The nest on the island I have mentioned was only a few yards from the bank of the river, along which persons were constantly passing, and so frequent were the attacks of the male bird that three hurdles drawn with boughs had to be placed as a screen as soon as the female began her nest; and this protection seemed to be fully appreciated by both birds.

Some years ago, one of the younger members of the family at Thoresby wished to obtain a swan's egg, and an old man whom I knew well, and who was a labourer in the grounds, was commissioned to procure one. A nest at the foot of a tree, in the pleasure grounds, contained several eggs, and old Thomas watched his opportunity, when the female was absent, to accomplish his task. Taking his garden rake in his hand he was about to secure his prize, when he was suddenly attacked by the female, who had seen his approach, and who at once flew up and felled him to the ground. Recovering his feet, he attempted to defend himself with his rake, was again knocked down, and, judging discretion the better part of valour, he beat a hasty retreat, leaving the parent bird in triumphant possession of her eggs.

The number of cygnets reared by each pair is generally from two to four; I have known six on two occasions, and once seven, but the latter number is very unusual.

Of the numerous family of the Anatidæ which I have seen in Sherwood Forest, few are constant residents with us; some, indeed, are very regular in their visits, while others I have to mention occur only occasionally. Amongst the latter is the Shieldrake (*Tadorna Bellonnii*), which, though a maritime species, I have seen twice on Thoresby Lake. It is a handsome bird, and the distinct colours of its plumage make it a very striking object on the water. I have known of its occurrence also on the river Idle, at Retford, and I have heard that they have bred there.

The Shoveller (*Anas clypeata*) is another of our rare visitors. On the 24th of October, 1854, I saw a male of this species on the lake at Thoresby, apparently alone, and watched it for some time. It appeared rather restless, though occasionally feeding for a few minutes, and then resuming its watchful attitude; but it soon took its departure, and I did not see it again. A pair, male and female, was seen on the 23rd of April, 1857, on the decoy at Houghton, and the male, I regret to say, was shot by the keeper there; or, from the time of the year, they might very likely have remained and bred in that very secluded place.

The Wild Duck (*A. boschas*), I need hardly say, is a

common species, found on all our streams, and is particularly abundant on all the large sheets of water. I think I have mentioned that one side of the lake at Thoresby is skirted for more than half its length by a broad belt of plantation, which, being securely fenced in, preserves that margin of the water in strict seclusion, and this is constantly thronged by an immense number of wild ducks. Many an hour's enjoyment I have had, while, hid from their observation, I have watched their various gambols. Aided by my glass, I could note every movement unsuspected. Some would be busy snapping rapidly at the water spiders and other insects, or, with tail poised in air, would endeavour to reach the water-weed below; others, with head laid back, or with bill buried in their plumage, floated lazily along, enjoying the *dolce far niente.* Presently, an arrival of three or four from a distant flight would put the whole flock in commotion, while with loud quacks they appeared to be questioning the new comers, who perchance had brought intelligence of some fresh feeding ground; while another detachment would quietly take their departure. Now and then they would make up their minds to a boisterous game of play, chasing one another with great vigour, and diving incessantly. But idle or busy, they were ever on the alert, and at the slightest symptom of danger the whole flock would sail towards the centre of the lake; or if the danger appeared more imminent, they at once with loud cries of alarm took their flight to a safer spot.

In severe winters, when the lakes are frozen, the large flocks of wild ducks, which had made them their homes during the rest of the year, disperse themselves in all directions, taking up their quarters on the running streams; yet at such seasons they often suffer severely from want, and become quite emaciated if the frost is of long continuance, losing much of their usual shyness and resorting to the fields.

A site for her nest is by no means invariably selected by the wild duck in the vicinity of water. The long heath on our open forest is constantly chosen for that purpose, and numberless are the nests I have known where the nearest stream has been at least a mile distant. How the young, when first hatched, were conveyed to the water, long surprised me; that it is done as soon as they break the shell I have no doubt, for though I have often found the eggs before hatching, and the empty shells after, I never met with the young ones. The extraordinary fact that they occasionally placed their nests in trees convinced me that the parent birds must in such cases carry their young at least to the ground; and I do not now doubt that this is their common practice where the distance to water is too great for the young ones to travel on foot.

I knew a nest which was placed in an evergreen in the pleasure grounds at Thoresby, and from which the young were hatched and brought safely off. Another instance came under my notice in 1856, in which the nest was constructed in a large beech tree, in an avenue in the same grounds, at the height of forty feet from the ground, and from this great elevation the young ones were safely conveyed to water.

The late Mr. Mansell of Thoresby related to me the following instance of the parent bird actually conveying her young to water, which he himself witnessed. He was passing one morning at daybreak, in the early part of May, under a large ash tree, which was thickly clothed with ivy, when a cheeping and rustling overhead induced him to withdraw a few steps and stand still. He had hardly done so when a wild duck flew out of the ivy, some height up the tree, holding a young one in her bill; this she put down on the bank of the stream, which was about a hundred yards from the tree, and then, returning to the nest, conveyed the remainder, one by one, in the same manner until thirteen were safely placed on the bank. Here she brooded them for a few minutes, and then with much apparent fondness led them down the bank into the water, where they were speedily darting about with the utmost liveliness.

This incident clearly illustrates how the difficulty of conveying the young from a height, or from a distance, is overcome, and I have little doubt of its being the common practice.

A large piece of water at Houghton, known as the "Decoy," was many years ago used as such, and a great number of wild ducks were annually captured there; but it has long fallen into disuse.

In May, 1855, I accidentally met with the nest of the Garganey (*A. querquedula*) in a small patch of furze, at no great distance from water. I did not see the birds, nor was I aware that they visited us. The nest was formed of dry grass, with a thick lining of down, and contained three addled eggs, and the broken shells of the others, which had been successfully hatched, their buff colour and size rendering them readily distinguishable from any allied species. My identification of them was confirmed by an experienced collector, who was with me at the time.

The Teal (*A. crecca*) is a constant winter visitor, occurring rather numerously on Thoresby Lake. The sheltered margin is a favourite haunt, and there in the daytime they float lazily about in large parties, betaking themselves at dusk to their feeding grounds. They generally arrive at the end of September or beginning of October, if the season is open, and take their departure about the middle of March; but I have seen stragglers a month later.

The Widgeon (*A. Penelope*) arrives about the same time as the teal, and frequents the same localities, but it is scarcely so numerous. I have often met with it on our streams, while the teal confines itself more strictly to the lakes, and have frequently heard their whistling cry while passing overhead in an evening, when I could not distinguish them. I am not aware of any instance of either of the last-named species breeding with us.

I must now pass over a number of species which are either but rare visitors to this country, or are strictly maritime in their habits—and in consequence are not to be expected to be found in my list—and come to that prettily marked duck the Red-headed Pochard (*A. ferina*), a few specimens of which are seen every winter. I cannot call it a common bird, though, as it is partial to quiet shaded streams in preference to more open waters, it perhaps escapes observation; a large number, however, frequented Thoresby Lake in the winter of 1860, and many were killed on the stream at New England, and Houghton, and other places.

Thoresby Lake is visited with great regularity every winter by large flocks of the Tufted Duck (*A. fuligula*). This is a small but lively species, constantly diving, but exceedingly shy and wary in its habits, and difficult of approach. In some years it is very numerous, as, for instance, in 1854, when I counted above one hundred together at a time, while in the winter of the following year immense flocks visited us. A few stragglers are occasionally found on the brooks, and in February, 1848, I saw a male bird for several successive days on the stream at the bottom of my own garden. Some floating rubbish had accumulated against a wooden grating, which stretched across the stream, and here the bird was feeding; but on my sudden approach it immediately took wing, uttering its alarm cry, which greatly resembled the wild duck's, but considerably shriller. On each of the following days I succeeded, under cover of a hedge, in getting within three or four yards, where I could watch it unperceived, and was greatly interested in its exceedingly lively active habits; but a very slight sound was sufficient to awaken its alarm, causing it to look round with a glance of its quick eye, erect its pendant crest, and take to flight if the sound was repeated.

I was one day watching a large flock of this species on Thoresby Lake, amused by their incessant diving and their active chases after each other, when I witnessed a singular freak on the part of a female, of turning herself over in the water and floating on her back for several minutes, while with her feet she appeared to be preening the feathers of the belly.

Towards the beginning of April the tufted ducks commence leaving us for the north, not taking their departure all at once, but gradually diminishing their numbers. In some years their stay is more prolonged than in others—doubtless influenced by the weather; for in 1856, when the spring was unusually cold and uncongenial, they remained with us until May 13, on which day I saw a great number engaged as usual in diving incessantly, but the following morning they had disappeared.

One or two pairs have occasionally abandoned their northern visit and remained with us to breed, choosing the long grass and rushes on the banks of the lake in which to place their nest. They have also been known to breed at Osberton.

In 1849 a small party visited a large pond in Colwick Park, near Nottingham, where they remained for several weeks, mingling in a friendly manner with some domestic ducks, and evincing little shyness, although the pond was close to the house.

The Golden-eyed Garrot (*A. clangula*) is generally amongst our winter visitors, though we cannot reckon on its regular arrival. In the winter of 1860 six or seven individuals frequented Thoresby Lake, and remained a few weeks; and through the winter of 1863 they were unusually plentiful.

Another handsome stranger is the Smew (*Mergus albellus*), which occasionally favours us with a visit. Several frequented the same water for a short time in February, 1855; and again, in 1860, I saw a pair on the same day I noticed the golden-eye. Five or six were also seen on the Trent in the winter of 1349, and several of these were shot; but it is only in hard winters that they come so far inland.

The smew shows wonderful activity in the water, diving on the slightest alarm; and I have been much astonished at the immense distance which it will pass under water.

Unlike the smew, the Goosander (*M. merganser*) regularly frequents the lakes in the parks, and in some years in considerable numbers, arriving about the end of October, and generally leaving in March; but I have known them prolong their stay until April—once, indeed, in 1854, I noted a flock of eighteen on Thoresby Lake so late as May 1.

The plumage of both the male and female goosander is very handsome, though so different that it is no wonder that for a long time, as in the similar case of the hen-harrier and ringtail, they were classed as distinct species, the female being known as the dun diver, and receiving the specific name of *castor*. This confusion of the sexes has, however, long since been cleared up by dissection, the anatomical peculiarities of the trachea being alike in both. I have myself frequently noticed amorous passages to take place between the goosander and the dun diver whilst on the water.

The contrast in the plumage of the male, between the black of the head and back and the rosy cream-colour of the neck and breast, is exceedingly striking, and though less obtrusive, the reddish-brown of the head and neck of the female, with the bluish-grey of the back, offers almost as pleasing a distinction.

The plumage of the young males is so exactly the same as that of the female, that until after their first moult it is impossible to distinguish them; yet they do not assume the full adult livery at once, as I have seen several in which the dark feathers of the head were interspersed with brown ones, the lower part of the neck and the breast being also mottled with ash colour. The pendant crest of both males and females adds greatly to their beauty; that of the former is looser in texture, and hangs more gracefully than that of the female, which, though of equal length, is narrower, and thicker at the upper part.

The numbers that frequent Thoresby Lake vary considerably: in some years there have not been more than fifteen or twenty; in others as many as thirty, or even forty, as in 1855. Welbeck Lake is generally visited by a small party, and sometimes Rufford also, and a few are annually seen on the Trent.

There is always a large proportion of males in adult plumage in every flock, and it is a very pretty sight to see them chasing each other, diving in apparent playfulness and emerging immediately; when they are feeding they remain much longer under water, and often traverse a space of seventy or eighty yards before coming to the surface. I have had many opportunities of observing their habits from the thick plantation on the border of the lake already mentioned, where I could conceal myself close to the water without being perceived, and many an hour have I passed there in watching their varied motions. They swim with their bodies low in the water, but yet with great activity and command, and I have seen them submerge themselves until their backs were almost covered; this is particularly the case with the female when making amorous advances to the male, her head and neck at the same time being outstretched and laid flat on the surface, until at a little distance she is almost invisible. Unaware of my vicinity, they often approached very near me, letting me inspect them closely. As the season drew on, the males gave some indications of choosing their partners; two or three would select the same female, and pursue her most unceremoniously for some time, she all the while diving incessantly to elude the pursuit of her obnoxious suitors, or to aid that of the favoured one, reminding me often of the racing courtships of the Tartar maidens and their mounted lovers. Towards the end of March their matrimonial engagements appeared to be concluded, and the flock was mostly broken up into pairs; but they became still more wary than before, and on the approach of any one near the opposite bank of the lake, which is there about a quarter of a mile wide, they would directly take wing to a greater distance, while in January they would merely swim slowly away.

I have never seen any of these on the land but once, and then it was only a single pair; their motions were, like those of any other waterfowl, rather ungainly, and greatly wanting in the activity they display on the water.

A single specimen of the Great Crested Grebe (*Podiceps cristatus*) was taken in April, 1856, under rather singular circumstances. It was a male bird, one of a pair which was seen by a labouring man on a small pond on the green, close to the village of Wellow. He was going to his work at dawn, and whilst crossing the green saw two strange birds on the pond busily diving; one of them dived out of sight on his approach, while the other with difficulty rose to the wing, but only flew a few yards before it fell to the ground, when the man ran up and secured it. As he had some distance to go to his work he did not remain to look after the other, which evidently was the female, and it was not seen again. It is somewhat remarkable that a small duck-pond in so public a place, and without the slightest shelter of reeds or rushes on its banks, or communication with any stream, should have been selected for a visit. I have not met with this species again.

The smallest of the family, the Little Grebe or dabchick (*P. minor*), is comparatively common in all our waters. At one time I thought it to be only resident with us in the summer, but I have found it plentiful in winter also, though apparently varying in numbers in different years. It constantly breeds with us in suitable spots, the nest being formed of a large quantity of flags and other water-plants, and those I have found have always been placed on the ground, close to the water, and not floating on that element. There is no difference between it and that of the coot—both are mere heaps of material, with a shallow cavity in the centre, and are very far removed from being *hot*-beds, as some have supposed they are. The eggs are oval, tapering to both ends, but to one rather more acutely than the other, and when fresh-laid are beautifully white; it is rarely, however, that they are found so, for whether with the wet feet of the bird, or the damp condition of the nest, they soon become sullied, until those first laid acquire an almost uniform brown hue, each successive egg being a shade lighter than the other.

No bird that I know has for its size greater power in the water than the dabchick; indeed, as a diver few surpass it. Heels over head it goes with the least possible splash, and if you could look into the water in which it has just dived you would perceive it gliding about with as easy a motion as a fish, using its wings to assist its progress, and seeming without any effort to keep below the surface. It is astonishing too how long it will remain beneath, and how rapidly it again disappears after it has come up to take breath. I once had an opportunity of witnessing these subaqueous gambols in a deep sluggish stream at the bottom of my garden, where I could look down into the water; but it is rarely that such an opportunity is afforded. So quickly do they dive, that I have shot at them at a distance of twenty yards more than once, and before the shot touched the water they had vanished.

Far away from its usual haunts on the ocean, a solitary specimen of the Black-throated Diver (*Colymbus arcticus*) was taken in the neighbourhood in January, 1848. It had alighted on the ice which covered a large piece of water, on which some snow had fallen and had partly thawed, but freezing again quickly in the evening, the poor bird was unwittingly detained a prisoner, and was found in the morning frozen to the ice and much exhausted. It was killed with a stick by the man who found it, and proved to be a female in good plumage.

In the winter of 1855 another dweller on the deep paid us a visit. This was the Common Guillemot (*Uria troile*), several of which frequented Thoresby Lake for a week or two in December of that year. They occupied themselves busily with fishing while they remained, and then suddenly took their departure.

The eggs of the common guillemot vary much in colour and markings, the most common ground colour being green, or dirty white, with streaks, spots, or blotches of dark reddish brown, or in some cases nearly black. I obtained one at Flamborough Head of a perfectly pure white, and I have one in my collection with the ground of a uniform warm buff, blotched as usual with brown.

During the severe frost of January, 1847, a specimen of the Little Auk (*Uria alle*) came into my hands. It was seen by a labouring man in a ditch on the banks of the Trent at Holme Pierrepoint, and as he thought it a "very strange sort of thing," as he afterwards described

it, he threw a hayfork, which he had in his hand at the
time, at the bird, and one of the prongs entering its
neck, he secured his prize without further trouble. The
little bird was in good flesh and plumage, but from its
making no attempt to escape when first seen, I imagine
it was exhausted with its southern journey. A few
days after this a Razorbill (*Alca tordu*) was seen on the
Trent at Nottingham, and was shot, the severe weather
having forced both these out of their usual course.

I have often wondered that attractions like ours have
not brought the Cormorant (*Carbo cormoranus*) more
into our neighbourhood. I am only aware of one, and
that was on Thoresby Lake, and it was soon shot. This
was in August, 1864. The swans regarded the stranger
with evident dislike, and chased it whenever he came
near.

The Shag (*Carbo cristatus*) I have known occur once,
a pair in immature plumage having been shot on the
Trent, at Burton Joyce, in the summer of 1851.

Although greatly circumscribed in its habitat, yet the
Gannet (*Sula bassana*) is sometimes driven from its
ocean home and found far inland. In 1837 a male bird,
in mature plumage, was seen on the lake in Welbeck
Park, and shot at, but being only injured in one wing,
it was captured and kept alive for some time on a small
pond. Here, while it lived, it fiercely defended its
limited dominions against all comers, and with unvarying success, for no intruders ventured to face his sharply
pointed bill. In 1849 three others were seen together
at Hexgrave, a few miles from Ollerton. They were
resting in a field, and permitted themselves to be
approached without difficulty, when one of them was

shot by the person who discovered them; the other two, on the death of their companion, immediately took to flight, and were not seen again.

Several members of the graceful family of the terns have frequently been noticed in our neighbourhood. A pair of the Common Tern (*Sterna hirundo*) visited our village on August 12th, 1857, and were flying up and down the stream, as if hawking for insects, at times skimming close to the water, and then rising again with all that elegant ease of motion and command of wing for which they are noted. But alas! their visit was cut short by the miller who lived close by, who shot one of them, which proved to be a male, and the other, which doubtless was its mate, was not seen again.

The Lesser Tern (*S. minuta*) has been seen several times on Thoresby Lake, and once or twice the Black Tern (*S. nigra*) has been in company. One of the latter species was shot out of a flock which was seen on the Trent, near Nottingham, in June, 1851; and, singularly enough, several made their appearance at the little village of Wilford, on the Trent (the scene of my schoolboy days), during a severe snowstorm in January, 1854.

I have noted four species of gulls which have been seen at various times and in various parts of our district, some of them occurring not unfrequently.

A Black-headed Gull (*Larus ridibundus*) was shot by a boy at the village of Boughton, two miles from Ollerton, on June 2nd, 1854, and was brought to me. It was a male in nearly mature plumage, the dark brown on the head and upper part of the neck being quite perfect, but the scapulars and wing coverts retained some slight mottling of brown, and the tail its dusky tips, except the centre, and one outer feather, which were pure white. There are several large colonies of these birds in Yorkshire, and this most likely was a straggler from one of them.

The Kittiwake (*L. tridactylus*) is frequently seen on the Trent, and I have noted its occurrence several times in our own immediate neighbourhood. On February 12th, 1850, I saw a young one busily searching some dung on the turnpike-road on Budby Forest; it rose on my approach, but soon alighted again, showing little sign of fear or timidity. On February 7th, 1854, a young bird in immature plumage was brought to me. It was first seen by the side of the race at Rufford Mill by the miller, and was evidently in an exhausted state, most likely from want of food during the severe weather that then prevailed, for he ran it down, after a short chase; it bore no marks of hurt or disease, and its plumage was in beautiful condition. On December 15th, 1857, two adult males were picked up, one at Southwell with its wing broken by shot, and the other at Farnsfield, dead. Another adult male was seen flying over the water at Perlethorpe Mill, and was shot by one of the keepers.

The Common Gull (*L. canus*) is a still more regular visitor, sometimes appearing in large flocks, but always in the winter. In January, 1848, during a severe frost, I saw a numerous party several times, and obtained one of them. In subsequent years I have frequently noticed their occurrence in greater or less abundance; but on each occasion they flew listlessly about, as if the want of food had deprived them of their usual busy energy; indeed, in November, 1859, one or two were picked up in quite an emaciated state. It seems strange that they should leave their usual marine feeding grounds at all, and especially to penetrate so far inland, when their prospects of congenial food are so uncertain.

The Lesser Black-backed Gull (*L. fuscus*) I have known to visit us twice, both times in May—once in 1855, at Bothamsal, and the other in 1859 at Markham Moor; both were young birds in immature plumage, and both were *of course* shot.

I have only one more to add to my list, and that is that little, active ocean roamer, the Stormy Petrel (*Thalassidroma pelagica*). It is strange that a bird so peculiarly maritime in its habits should have been noted in the heart of Sherwood Forest! Few birds visit *terra firma* less than this smallest of our sea birds; and those that have been taken in various parts of England have, I believe, always been in winter, and have generally been blown out of their marine haunts by gales or storms.

A pair, male and female, was shot on Thoresby Lake in the winter of 1845, and thus I claim a place for them in our local fauna—they were skimming over the water in their usual manner.

Often have I with delight watched the ceaseless activity of these little dwellers on the sea; whether the sea was calm, or whether

> "In breeze, or gale, or storm,
> Icing the pole, or in the torrid zone dark heaving,"

it made no matter—there was this little petrel steadily pursuing its way. The boundless ocean is its home, and I have seen it more than a thousand miles from the nearest land, tipping the waves with its little feet, and following in the wake of the ship to pick up such fragments as might be thrown overboard. I have often amused myself by dropping bits of biscuit, and watching the race that was made for them by our numerous followers; sometimes the powerful albatross would glide by on outstretched and motionless wings and snatch up the morsel; but oftener the little stormy petrel succeeded in picking it up, though sometimes its right would be disputed by one of its more powerful neighbours. It seems to possess more discrimination too than some of the latter; for though I have captured the albatross and some of the larger sea birds by the common method of baiting a fish-hook with a bit of pork, I never succeeded in hooking a storm petrel, though they constantly followed in our wake.

My task is done, and I lay down my pen with some regret, for I have in thought lived over many pleasant hours, and revisited scenes to which I am much attached. I only hope my readers have derived pleasure from my scanty notes, and that some may thereby be induced to undertake for their own localities, what I have very imperfectly attempted for Sherwood Forest.

APPENDIX.

BILL OF CROSSBILL, p. 124.

This bird so seldom breeds in this country, that opportunities for examination are very rare, and in reference to the crossing of the mandibles, since the text was written I have met with a notice by Mr. Blyth, of a nest of the common species that was taken in the vicinity of Sevenoaks. The young birds, four in number, were half fledged when found, and were uniformly brown in colour; but the most noticeable peculiarity was that the mandibles of each were as much crossed as those of the adults. This, therefore, confirms the opinion I have expressed that this is their normal form. The straight mandibles of the young one mentioned by Yarrell must be considered an exception, and indeed were literally deformed.

NUTHATCH, p. 150.

I have lately noticed a peculiarity in the mode in which the Nuthatch gives the heavy blows necessary to penetrate the thick-shelled nuts. The bill is particularly thick and strong for so small a bird, and approaches in character the wedge-shaped bill of the woodpeckers. Formidable as this weapon is, its efficiency is increased by the singular manner in which it is sometimes used. The bird seems to have an idea when additional force is required, and instead of striking with the bill by the movement of the neck alone, it throws the whole weight of its body into the blow, turning itself as it were into the head of a hammer, which swings upon the feet as a pivot.

APPENDIX.

Oil Glands of Birds, p. 213.

In investigating the question of the use of the oil-glands in birds, I examined a large number of species for the purpose of ascertaining how these glands were distributed, some possessing only one, others two. In this I have been greatly assisted by Mr. Dunn, whose extensive opportunities as a collector have rendered his kind help extremely valuable. The following are the species examined:

1. *Those possessing one Gland.*

Griffon Vulture.
Egyptian Vulture.
Golden Eagle.
Spotted Eagle.
White-tailed Eagle.
Osprey.
Iceland Falcon.
Peregrine Falcon.
Hobby.
Merlin.
Red-footed Falcon.
Kestrel.
Goshawk.
Sparrowhawk.
Kite.
Buzzard.
Rough-legged Buzzard.
Honey Buzzard.
Marsh Harrier.
Hen Harrier.
Ashcoloured Harrier.
Eagle Owl.
Scops-Eared Owl.
Long-Eared Owl.
Short-Eared Owl.
Barn Owl.
Tawny Owl.
Snowy Owl.
Hawk Owl.
Little Owl.
Great Grey Shrike.
Red-backed Shrike.
Woodchat Shrike.
Fieldfare.
Song Thrush.
Rock Thrush.
Redwing.
Blackbird.
Ring Ousel.
Golden Oriole.
Alpine Accentor.
Hedge Sparrow.
Redbreast.
Blue-throated Warbler.
Redstart.
Black Redstart.
Wheatear.
Great Sedge Warbler.

APPENDIX.

Orpheus Warbler.
Rufous Sedge Warbler.
Melodious Willow Wren.
Dartford Warbler.
Great Titmouse.
Blue Titmouse.
Crested Titmouse.
Marsh Titmouse.
Bearded Titmouse.
Bohemian Waxwing.
White Wagtail.
Pied Wagtail.
Grey Wagtail.
Grey-headed Wagtail.
Ray's Wagtail.
Meadow Pipit.
Rock Pipit.
Sky Lark.
Wood Lark.
Crested Lark.
Short-toed Lark.
Shore Lark.
Snow Bunting.
Lapland Bunting.
Common Bunting.
Black-headed Bunting.
Yellowhammer.
Cirl Bunting.
Ortolan Bunting.
Chaffinch.
Brambling.
Tree Sparrow.
House Sparrow.
Greenfinch.
Goldfinch.
Linnet.
Lesser Redpoll.

Mealy Redpoll.
Twite.
Crossbill.
Parrot Crossbill.
Starling.
Chough.
Raven.
Crow.
Hooded Crow.
Rook.
Jackdaw.
Magpie.
Jay.
Nutcracker.
Great Black Woodpecker.
Green Woodpecker.
Spotted Woodpecker.
Lesser Spotted Woodpecker.
Wryneck.
Wren.
Roller.
Swallow.
Purple Martin.
White-bellied Swift.
Rock Dove.
Turtle Dove.
Passenger Pigeon.
Pheasant.
Capercailzie.
Black Grouse (small).
Partridge.
Barbary Partridge.
Quail.
Andalusian Quail.
Virginian Colin.

2. *Those with two Glands.*

Dipper.
Belted Kingfisher.
Kingfisher.
Red Grouse (small).
Ptarmigan (small).
Golden Plover.
Dotterell.
Ringed Plover.
Kentish Plover.
Little Ringed Plover.
Gray Plover.
Lapwing.
Turnstone.
Sanderling.
Oyster Catcher.
Heron.
Purple Heron.
Great White Heron.
Buff-backed Heron.
Squacco Heron.
Bittern.
Little Bittern.
American Bittern.
Night Heron.
White Stork.
Spoonbill.
Glossy Ibis.
Curlew.
Whimbrel.
Redshank.
Green Sandpiper.
Wood Sandpiper.
Common Sandpiper.
Spotted Sandpiper.
Greenshank.
Bartram's Sandpiper.
Black-winged Stilt.
Ruff.
Woodcock.
Brown Snipe.
Knot.
Temminck's Stint.
Dunlin.
Purple Sandpiper.
Spotted Crake.
Little Crake.
Baillon's Crake.
Moor Hen.
Water Rail.
Coot.
Grey Phalarope.
Red-necked Phalarope.
Gray Lag Goose.
Bean Goose.
Pink-footed Goose.
White-fronted Goose.
Bernicle Goose.
Brent Goose.
Canada Goose.
Hooper.
Ruddy Shieldrake.
Common Shieldrake.
Shoveller.
Gadwall.
Pintailed Duck.
Wild Duck.
Garganey.
Teal.

APPENDIX.

Widgeon.
American Widgeon.
Eider Duck.
Velvet Scoter.
Common Scoter.
Surf Scoter.
Long-tailed Duck.
Pochard.
Ferruginous Duck.
Scaup Duck.
Tufted Duck.
Harlequin Duck.
Goldeneye.
Buffel-headed Duck.
Smew.
Hooded Merganser.
Red-breasted Merganser.
Goosander.
Red-necked Grebe.
Sclavonian Grebe (very large).
Eared Grebe.
Little Grebe.
Great Northern Diver (large).
Black-throated Diver.
Red-throated Diver (large).
Guillemot.
Brunnich's Guillemot.
Ringed Guillemot.
Black Guillemot.
Little Auk.

Puffin.
Razorbill.
Cormorant.
Shag.
Gannet.
Sandwich Tern.
Roseate Tern.
Common Tern.
Arctic Tern.
Whiskered Tern.
Gull-billed Tern.
Sabine's Gull.
Little Gull.
Black-headed Gull.
Laughing Gull.
Kittiwake.
Common Gull.
Iceland Gull.
Lesser Black-backed Gull.
Herring Gull.
Great Black-backed Gull.
Glaucous Gull.
Common Skua.
Pomarine Skua.
Richardson's Skua.
Buffon's Skua.
Fulmar Petrel.
Great Shearwater.
Manx Shearwater.
Fork-tailed Petrel.
Storm Petrel.

INDEX.

	PAGE
AUK, Little	228
Avocet	193
BAR-TAILED Godwit	194
Bittern	192
Black Grouse	184
Black Tern	230
Black-headed Gull	230
Black-tailed Godwit	194
Black-throated Diver	228
Black-winged Stilt	194
Blackbird	58
Blackcap	77
Blackstart	67
Blue Tit	83
Bohemian Chatterer	88
Brambling	103
Bullfinch	121
Bunting, Black-headed	99
,, Cirl	101
,, Common	98
,, Snow	98
,, Yellow	100
Bustard, Little	187
Buzzard, Common	31
,, Rough-legged	31
,, Honey	32
CHIFFCHAFF	81
Cirl Bunting	101
Chaffinch	101
Clipstone Palace	1
Cole Tit	85
Common Bunting	98
,, Gull	231
,, Tern	230

	PAGE
Coot	202
Cormorant	229
Crake, Spotted	200
Creeper	147
Crossbill	123
,, Parrot	125
,, White-winged	126
Crow, Carrion	129
,, Hooded	129
Cuckoo	151
Curlew	192
DABCHICK	226
Dipper	50
Diver, Black-throated	228
Dove, Ring	176
,, Stock	179
,, Turtle	179
Duck, Tufted	221
,, Wild	217
EAGLE, Golden	18
,, White-tailed	18
FIELDFARE	55
Finch, Gold	117
,, Green	116
,, Haw	116
,, Mountain	103
,, Siskin	118
Flycatcher, Pied	49
,, Spotted	49
GARDEN Warbler	77
Garganey	220

R

INDEX.

	PAGE
Gannet	229
Godwit, Bar-tailed	194
,, Black-tailed	194
Goldcrest	82
Golden Plover	188
Goldeneye	223
Goldfinch	117
Goosander	223
Goose, Bean	214
,, Grey-lag	213
Goshawk	29
Grasshopper Warbler	71
Great Crested Grebe	226
Great Tit	83
Grebe, Great Crested	226
,, Little ,,	226
Green Woodpecker	143
Greendale Oak	7
Greenfinch	116
Greenshank	193
Grey Plover	189
Grey Wagtail	91
Grey-headed Wagtail	92
Grouse, Black	184
,, Red	184
Guillemot, Common	228
Gull, Black-headed	230
,, Common	231
,, Kittiwake	231
,, Lesser Black-backed	232

	PAGE
HARRIER, Hen	32
,, Marsh	32
Hawfinch	116
Hedge Sparrow	61
Heron	189
Hobby	24
Hooded Crow	129
Hooper	214
Hoopoe	149
House Martin	165
,, Sparrow	104
Hybrid, Game	182

	PAGE
INOSCULATED Trees	10

	PAGE
JACKDAW	138
Jack Snipe	198
Jay	142

	PAGE
KESTREL	26
Kingfisher	159
Kite	29
Kittiwake Gull	231

	PAGE
LANDRAIL	198
Lapwing	189
Lark, Sky	95
,, Wood	97
Lesser Black-backed Gull	232
,, Redpole	120
,, Spotted Woodpecker	146
,, Tern	230
,, Whitethroat	79
Linnet	120
,, Lesser Redpole	120
,, Mountain	121
Little Auk	228
,, Bustard	187
,, Grebe	226
Long-tailed Tit	86

	PAGE
MAGPIE	140
Major Oak	9
Marsh, Tit	86
Martin, House	165
,, Sand	168
Meadow Pipit	95
Merlin	25
Mice, destruction of trees by	38
Mountain Finch	103
,, Linnet	121
Mute Swan	214

	PAGE
NEWSTEAD ABBEY	2
Nightingale	74
Nightjar	171
Nuthatch	149

	PAGE
OIL Glands	205
Osprey	20
Ousel, Ring	60
,, Water	50
Owl, Eagle	39
,, Long-eared	39
,, Note of	43
,, Short-eared	41

INDEX. 243

	PAGE
Owl, Tawny	44
,, White	41

	PAGE
PARLIAMENT Oak	6
Parrot, Crossbill	125
Partridge	184
,, Red-legged	186
Peregrine Falcon	21
Petrel, Stormy	232
Pheasant	180
Pied Flycatcher	49
,, Wagtail	89
Pipit, Meadow	95
,, Tree	93
Plover, Common	189
,, Golden	188
,, Grey	189
,, Stone	188
Pochard, Red-headed	221

	PAGE
QUAIL	186

	PAGE
RAVEN	128
Razorbill	229
Red Grouse	184
Red-headed Pochard	221
Red-legged Partridge	186
Redshank	193
Redstart	65
,, Black	67
Redwing	56
Ring Dove	176
Ringtail	33
Robin	62
Rook	132
Rufford Abbey	2

	PAGE
SAND Martin	168
Sandpiper, Common	193
Sedge Warbler	72
Shag	229
Shambles Oak	9
Shieldrake	217
Shoveller	217
Shrike, Great Grey	46
,, Red-backed	46

	PAGE
Shrike, Wood Chat	48
Siskin	118
Skylark	95
Smew	223
Snipe, Common	197
,, Jack	198
Snow Bunting	98
Sparrow, Hedge	61
,, House	104
,, Tree	103
Sparrowhawk	27
Spotted Crake	200
,, Woodpecker	145
Starling	126
Stilt, Black-winged	194
Stock Dove	179
Stonechat	69
Stormy Petrel	232
Swallow	160
Swan, Mute	214
,, Hooper	214
Swift	169

	PAGE
TEAL	220
Tern, Black	230
,, Common	230
,, Lesser	230
Thick-knee	188
Thrush, Missel	51
,, Song	56
Titmouse, Blue	83
,, Cole	85
,, Great	83
,, Long-tailed	86
,, Marsh	86
Tree Pipit	93
,, Sparrow	103
Turtle Dove	179
Tufted Duck	221
Twite	121

	PAGE
WAGTAIL	88
,, Grey	91
,, Grey-headed	92
,, Pied	89
,, White	91
,, Yellow	92
Warbler, Garden	77
,, Grasshopper	71

INDEX.

	PAGE
Warbler, Reed	74
,, Sedge	72
Water Hen	200
,, Rail	201
Waxwing	88
Welbeck Abbey	2
Wheatear	70
Whinchat	70
White Wagtail	91
White-winged Crossbill	126
Whitethroat	78
,, Lesser	79
Widgeon	221
Wild Duck	217
Willow Wren	81
Wood Lark	97
,, Wren	80
Woodchat	48
Woodcock	194
Woodpecker, Green	143
,, Lesser, Spotted	146
,, Spotted	145
Wren	147
,, Gold-crested	82
Wryneck	146
YELLOW Bunting	100
,, Wagtail	92
,, Willow Wren	81

THE END.

LIST OF WORKS

PUBLISHED BY

L. REEVE & CO.

NEW SERIES OF NATURAL HISTORY FOR BEGINNERS.

₊ A good introductory series of books on Natural History for the use of students and amateurs is still a *desideratum*. Those at present in use have been too much compiled from antiquated sources; whilst the figures, copied in many instances from sources equally antiquated, are far from accurate, the colouring of them having become degenerated through the adoption, for the sake of cheapness, of mechanical processes.

The present series will be entirely the result of original research carried to its most advanced point; and the figures, which will be chiefly engraved on steel, by the artist most highly renowned in each department for his technical knowledge of the subjects, will in all cases be drawn from actual specimens, and coloured separately by hand.

Each work will treat of a department of Natural History sufficiently limited in extent to admit of a satisfactory degree of completeness.

The following are now ready:—

British Insects; a Familiar Description of the Form, Structure, Habits, and Transformations of Insects. By E. F. STAVELEY. Crown 8vo, 16 Coloured Steel Plates, engraved from Natural Specimens expressly for the work by E. W. ROBINSON, and numerous Wood-Engravings by E. C. RYE, 14s.

British Butterflies and Moths; an Introduction to the Study of our Native LEPIDOPTERA. By H. T. STAINTON. Crown 8vo, 16 Coloured Steel Plates, containing Figures of 100 Species, engraved from Natural Specimens expressly for the work by E. W. ROBINSON, and Wood-Engravings, 10s. 6d.

British Beetles; an Introduction to the Study
of our Indigenous COLEOPTERA. By E. C. RYE. Crown 8vo, 16 Coloured Steel Plates, comprising Figures of nearly 100 Species, engraved from Natural Specimens, expressly for the work, by E. W. ROBINSON, and 11 Wood-Engravings of Dissections by the Author, 10s. 6d.

British Bees; an Introduction to the Study of
the Natural History and Economy of the Bees indigenous to the British Isles. By W. E. SHUCKARD. Crown 8vo, 16 Coloured Steel Plates, containing nearly 100 Figures, engraved from Natural Specimens, expressly for the work, by E. W. ROBINSON, and Woodcuts of Dissections, 10s. 6d.

British Spiders; an Introduction to the Study
of the ARANEIDÆ found in Great Britain and Ireland. By E. F. STAVELEY. Crown 8vo, 16 Plates, containing Coloured Figures of nearly 100 Species, and 40 Diagrams, showing the number and position of the eyes in various Genera, drawn expressly for the work by TUFFEN WEST, and 44 Wood-Engravings, 10s. 6d.

British Grasses; an Introduction to the Study
of the Grasses found in the British Isles. By M. PLUES. Crown 8vo, 16 Coloured Plates, drawn expressly for the work by W. FITCH, and 100 Wood-Engravings, 10s. 6d.

British Ferns; an Introduction to the Study
of the FERNS, LYCOPODS, and EQUISETA indigenous to the British Isles. With Chapters on the Structure, Propagation, Cultivation, Diseases, Uses, Preservation, and Distribution of Ferns. By M. PLUES. Crown 8vo, 16 Coloured Plates, drawn expressly for the work by W. FITCH, and 55 Wood-Engravings, 10s. 6d.

British Seaweeds; an Introduction to the Study
of the Marine ALGÆ of Great Britain, Ireland, and the Channel Islands. By S. O. GRAY. Crown 8vo, 16 Coloured Plates, drawn expressly for the work by W. FITCH, 10s. 6d.

Other Works in preparation.

BOTANY.

The Young Collector's Handybook of Botany.
By the Rev. H. P. DUNSTER, M.A. 66 Wood-Engravings, 4s. 6d.

The Natural History of Plants. By H.
BAILLON, President of the Linnæan Society of Paris, Professor of Medical Natural History and Director of the Botanical Garden of the Faculty of Medicine of Paris. Translated by MARCUS M. HARTOG, Trinity College, Cambridge. Super-royal 8vo. Vol. I., with 503 Wood-Engravings, 25s. Also in Monthly Parts at 2s. 6d.

Domestic Botany; an Exposition of the
Structure and Classification of Plants, and of their uses for Food, Clothing, Medicine, and Manufacturing Purposes. By JOHN SMITH, A.L.S., ex-Curator of the Royal Botanic Gardens, Kew. Crown 8vo, 16 Coloured Plates and Wood-Engravings, 16s.

Handbook of the British Flora; a Description
of the Flowering Plants and Ferns indigenous to, or naturalized in, the British Isles. For the Use of Beginners and Amateurs. By GEORGE BENTHAM, F.R.S., President of the Linnæan Society. New Edition, Crown 8vo, 12s.

The Illustrated British Flora, a Description
(with a Wood-Engraving, including dissections, of each species) of the Flowering Plants and Ferns indigenous to, or naturalized in, the British Isles. By GEORGE BENTHAM, F.R.S., President of the Linnæan Society. Demy 8vo, 2 vols., 1295 Wood-Engravings, from Original Drawings by W. FITCH, £3 10s.

British Wild Flowers, Familiarly Described
in the Four Seasons. A New Edition of "The Field Botanist's Companion." By THOMAS MOORE, F.L.S. Demy 8vo, 24 Coloured Plates, by W. FITCH, 16s.

British Grasses; an Introduction to the Study of the Gramineæ of Great Britain and Ireland. By M. PLUES. Crown 8vo, 100 Wood-Engravings, 6s.; with 16 Coloured Plates by W. FITCH, 10s. 6d.

Curtis's Botanical Magazine: Figures and Descriptions of New and Rare Plants of Interest to the Botanical Student, and suitable for the Garden, Stove, or Greenhouse. By Dr. J. D. HOOKER, F.R.S., Director of the Royal Gardens, Kew. Royal 8vo. Published Monthly, with 6 Plates, 3s. 6d. coloured.

The Floral Magazine: Figures and Descriptions of Select New Flowers for the Garden, Stove, or Conservatory. By the Rev. H. HONYWOOD DOMBRAIN, A.B. New Series, enlarged to Royal 4to. Monthly, with 4 Plates, 3s. 6d. coloured.

Outlines of Elementary Botany, as Introductory to Local Floras. By GEORGE BENTHAM, F.R.S., President of the Linnæan Society. Second Edition, 2s. 6d.

The Tourist's Flora; a Descriptive Catalogue of the Flowering Plants and Ferns of the British Islands, France, Germany, Switzerland, Italy, and the Italian Islands. By JOSEPH WOODS, F.L.S. Demy 8vo, 18s.

Contributions to the Flora of Mentone, and to a Winter Flora of the Riviera, including the Coast from Marseilles to Genoa. By J. TRAHERNE MOGGRIDGE, F.L.S. Royal 8vo. In 4 parts, each, with 25 Coloured Plates, 15s., or complete in one vol. 63s.

Flora of Tasmania. By Dr. J. D. HOOKER, F.R.S. 2 vols. Royal 4to. 200 Plates, £17 10s., coloured. Published under the authority of the Lords Commissioners of the Admiralty.

Flora of Tropical Africa. By DANIEL
OLIVER, F.R.S., F.L.S. Vols I. and II., 20*s*. each. Published
under the authority of the First Commissioner of Her Majesty's
Works.

Handbook of the New Zealand Flora ; a
Systematic Description of the Native Plants of New Zealand,
and the Chatham, Kermadec's, Lord Auckland's, Campbell's,
and Macquarrie's Islands. By Dr. J. D. HOOKER, F.R.S.
Demy 8vo. Part I., 16*s*. ; Part II., 14*s*. ; or complete in one
vol., 30*s*. Published under the auspices of the Government of
that colony.

Flora Australiensis; a Description of the
Plants of the Australian Territory. By GEORGE BENTHAM,
F.R.S., President of the Linnæan Society, assisted by FERDI-
NAND MUELLER, F.R.S., Government Botanist, Melbourne,
Victoria. Demy 8vo. Vols. I. to V., 20*s*. each. Published
under the auspices of the several Governments of Australia.

Flora of the British West Indian Islands. By
Dr. GRISEBACH, F.L.S. Demy 8vo, 37*s*. 6*d*. Published under
the auspices of the Secretary of State for the Colonies.

Flora Hongkongensis; a Description of the
Flowering Plants and Ferns of the Island of Hongkong. By
GEORGE BENTHAM, P.L.S. With a Map of the Island. Demy
8vo, 16*s*. Published under the authority of Her Majesty's
Secretary of State for the Colonies.

Flora Vitiensis; a Description of the Plants
of the Viti or Fiji Islands, with an Account of their History,
Uses, and Properties. By Dr. BERTHOLD SEEMANN, F.L.S.
Royal 4to, Parts I. to IX. each, 10 Coloured Plates, 15*s*. To
be completed in 10 Parts.

On the Flora of Australia, its Origin, Affini-ties, and Distribution; being an Introductory Essay to the "Flora of Tasmania." By Dr. J. D. HOOKER, F.R.S., 10s.

Genera Plantarum, ad Exemplaria imprimis in Herbariis Kewensibus servata definita. By GEORGE BENTHAM, F.R.S., President of the Linnæan Society, and Dr. J. D. HOOKER, F.R.S., Director of the Royal Gardens, Kew. Vol. I. Part I. Royal 8vo, 21s. Part II., 14s.; Part III., 15s.; or Vol. I. complete, 50s.

Laws of Botanical Nomenclature adopted by the International Botanical Congress, with an Historical Introduction and a Commentary. By ALPHONSE DE CANDOLLE. 2s. 6d.

Illustrations of the Nueva Quinologia of Pavon, with Observations on the Barks described. By J. E. HOWARD, F.L.S. With 27 coloured Plates by W. FITCH. Imperial folio, half-morocco, gilt edges, 6l. 6s.

The Quinology of the East Indian Plantations. By J. E. HOWARD, F.L.S. Folio, 3 Coloured Plates, 21s.

Revision of the Natural Order Hederaceæ, being a reprint, with numerous additions and corrections, of a series of papers published in the "Journal of Botany, British and Foreign." By BERTHOLD SEEMANN, Ph.D., F.L.S. 8vo, 7 Plates, 10s. 6d.

Illustrations of the Genus Carex. By FRANCIS BOOTT, M.D. Part IV. Folio, 189 Plates, 10l.

Icones Plantarum. Figures, with Brief Descriptive Characters and Remarks, of New and Rare Plants, selected from the Author's Herbarium. By Sir W. J. HOOKER, F.R.S. New Series, Vol. V. 100 Plates, 31s. 6d.

A Second Century of Orchidaceous Plants,
selected from the Subjects published in Curtis's "Botanical Magazine" since the issue of the "First Century." Edited by JAMES BATEMAN, Esq., F.R.S. Complete in 1 Vol. royal 4to, 100 Coloured Plates, 5l. 5s.

Monograph of Odontoglossum, a Genus of the
Vandeous Section of Orchidaceous Plants. By JAMES BATEMAN, Esq., F.R.S. Imperial folio, Parts I. to IV., each with 5 Coloured Plates, and occasional Wood Engravings, 21s.

Select Orchidaceous Plants. By ROBERT
WARNER, F.R.H.S. With Notes on Culture by B. S. WILLIAMS. Folio, cloth gilt, 6l. 6s.
Second Series, Parts I. to VIII., each, with 3 Coloured Plates, 10s. 6d.

The Rhododendrons of Sikkim-Himalaya;
being an Account, Botanical and Geographical, of the Rhododendrons recently discovered in the Mountains of Eastern Himalaya from Drawings and Descriptions made on the spot, by Dr. J. D. Hooker, F.R.S. By Sir W. J. HOOKER, F.R.S. Folio, 30 Coloured Plates, 4l. 14s. 6d.

FERNS.

British Ferns; an Introduction to the Study
of the FERNS, LYCOPODS, and EQUISETA indigenous to the British Isles. With Chapters on the Structure, Propagation, Cultivation, Diseases, Uses, Preservation, and Distribution of Ferns. By M. PLUES. Crown 8vo, 55 Wood Engravings, 6s.; with 16 Coloured Plates by W. FITCH, 10s. 6d.

The British Ferns; Coloured Figures and Descriptions, with Analysis of the Fructification and Venation of the Ferns of Great Britain and Ireland, systematically arranged. By Sir W. J. HOOKER, F.R.S. Royal 8vo, 66 Coloured Plates, 2l. 2s.

GardenFerns; Coloured Figures and Descrip-
tions, with Analysis of the Fructification and Venation, of a Selection of Exotic Ferns, adapted for Cultivation in the Garden, Hothouse, and Conservatory. By Sir W. J. HOOKER, F.R.S. Royal 8vo, 64 Coloured Plates, 2l. 2s.

Filices Exoticæ; Coloured Figures and De-
scription of Exotic Ferns. By Sir W. J. HOOKER, F.R.S. Royal 4to, 100 Coloured Plates, 6l. 11s.

Ferny Combes; a Ramble after Ferns in the
Glens and Valleys of Devonshire. By CHARLOTTE CHANTER. *Third Edition.* Fcap. 8vo, 8 Coloured Plates by FITCH, and a Map of the County, 5s.

MOSSES.

Handbook of British Mosses, containing all
that are known to be natives of the British Isles. By the Rev. M. J. BERKELEY, M.A., F.L.S. Demy 8vo, 24 Coloured Plates, 21s.

SEAWEEDS.

British Seaweeds; an Introduction to the
Study of the Marine ALGÆ of Great Britain, Ireland, and the Channel Islands. By S. O. GRAY. Crown 8vo, 6s.; with 16 Coloured Plates, drawn expressly for the work by W. FITCH, 10s. 6d.

Phycologia Britannica; or, History of British
Seaweeds, containing Coloured Figures, Generic and Specific Characters, Synonyms and Descriptions of all the Species of Algæ inhabiting the Shores of the British Islands. By Dr. W. H. HARVEY, F.R.S. New Edition. Royal 8vo, 4 vols. 360 Coloured Plates, 7l. 10s.

Phycologia Australica; a History of Australian Seaweeds, comprising Coloured Figures and Descriptions of the more characteristic Marine Algæ of New South Wales, Victoria, Tasmania, South Australia and Western Australia, and a Synopsis of all known Australian Algæ. By Dr. HARVEY, F.R.S. Royal 8vo, 5 vols., 300 Coloured Plates, 7l. 13s.

Nereis Australis; or, Algæ of the Southern Ocean, being Figures and Descriptions of Marine Plants collected on the Shores of the Cape of Good Hope, the extra-tropical Australian Colonies, Tasmania, New Zealand, and the Antarctic Regions. By Dr. HARVEY, F.R.S. Imperial 8vo, 50 Coloured Plates, 2l. 2s.

FUNGI.

Outlines of British Fungology, containing Characters of above a Thousand Species of Fungi, and a Complete List of all that have been described as Natives of the British Isles. By the Rev. M. J. BERKELEY, M.A., F.L.S. Demy 8vo, 24 Coloured Plates, 30s.

The Esculent Funguses of England. Containing an Account of their Classical History, Uses, Characters, Development, Structure, Nutritious Properties, Modes of Cooking and Preserving, &c. By C. D. BADHAM, M.D. Second Edition. Edited by F. CURREY, F.R.S. Demy 8vo, 12 Coloured Plates, 12s.

Illustrations of British Mycology, comprising Figures and Descriptions of the Funguses of interest and novelty indigenous to Britain. By Mrs. T. J. HUSSEY. Royal 4to. Second Series, 50 Coloured Plates, £4. 10s.

Clavis Agaricinorum: an Analytical Key to the British Agaricini, with Characters of the Genera and Subgenera. By WORTHINGTON G. SMITH, F.L.S. Six Plates. 2s. 6d.

SHELLS AND MOLLUSKS.

Elements of Conchology; an Introduction to the Natural History of Shells, and of the Animals which form them. By LOVELL REEVE, F.L.S. Royal 8vo, 2 vols. 62 Coloured Plates, £2. 16s.

Conchologia Iconica; or, Figures and Descriptions of the Shells of Mollusks, with remarks on their Affinities, Synonymy, and Geographical Distribution. By LOVELL REEVE, F.L.S. Demy 4to, in double Parts, with 16 Coloured Plates. 20s.

Conchologia Indica; Illustrations of the Land and Freshwater Shells of British India. Edited by SYLVANUS HANLEY, F.L.S., and WILLIAM THEOBALD, of the Geological Survey of India. 4to, Parts I. to III., each with 20 Coloured Plates, 20s.

The Edible Mollusks of Great Britain and Ireland, with the Modes of Cooking them. By M. S. LOVELL. Crown 8vo, 5s.; with 12 Coloured Plates, 8s. 6d.

INSECTS.

British Insects. A Familiar Description of the Form, Structure, Habits, and Transformations of Insects. By E. F. STAVELEY, Author of "British Spiders." Crown 8vo, with 16 beautifully Coloured Steel Plates and numerous Wood-Engravings, 14s.

British Beetles; an Introduction to the Study of our Indigenous COLEOPTERA. By E. C. RYE. Crown 8vo, 16 Coloured Steel Plates, comprising Figures of nearly 100 Species, engraved from Natural Specimens, expressly for the work, by E. W. ROBINSON, and 11 Wood-Engravings of Dissections by the Author, 10s. 6d.

British Bees; an Introduction to the Study
of the Natural History and Economy of the Bees Indigenous to
the British Isles. By W. E. SHUCKARD. Crown 8vo, 16 Coloured
Steel Plates, containing nearly 100 Figures, engraved from
Natural Specimens, expressly for the work, by E. W. ROBINSON, and Woodcuts of Dissections, 10s. 6d.

British Butterflies and Moths; an Introduc-
tion to the Study of our Native LEPIDOPTERA. By H. T.
STAINTON. Crown 8vo, 16 Coloured Steel Plates, containing
Figures of 100 Species, engraved from Natural Specimens expressly for the work by E. W. ROBINSON, and Wood-Engravings, 10s. 6d.

British Spiders; an Introduction to the Study
of the ARANEIDÆ found in Great Britain and Ireland. By E.
F. STAVELEY. Crown 8vo, 16 Plates, containing Coloured
Figures of nearly 100 Species, and 40 Diagrams, showing the
number and position of the eyes in various Genera, drawn expressly for the work by TUFFEN WEST, and 44 Wood-Engravings, 10s. 6d.

Curtis's British Entomology. Illustrations
and Descriptions of the Genera of Insects found in Great
Britain and Ireland, containing Coloured Figures, from Nature,
of the most rare and beautiful Species, and, in many instances,
upon the plants on which they are found. 8 vols. Royal 8vo,
770 coloured Plates, £21.

Or in separate Monographs.

Orders.	Plates.	£	s.	d.	Orders.	Plates.	£	s.	d.
APHANIPTERA	2	0	2	0	HYMENOPTERA	125	4	0	0
COLEOPTERA	256	8	0	0	LEPIDOPTERA	193	6	0	0
DERMAPTERA	1	0	1	0	NEUROPTERA	13	0	9	0
DICTYOPTERA	1	0	1	0	OMALOPTERA	6	0	4	6
DIPTERA	103	3	5	0	ORTHOPTERA	5	0	4	0
HEMIPTERA	32	1	1	0	STREPSIPTERA	3	0	2	6
HOMOPTERA	21	0	14	0	TRICHOPTERA	9	0	6	6

"Curtis's Entomology," which Cuvier pronounced to have "reached
the ultimatum of perfection," is still the standard work on the
Genera of British Insects. The Figures executed by the author
himself, with wonderful minuteness and accuracy, have never been
surpassed, even if equalled. The price at which the work was originally published was £43 16s.

Insecta Britannica; Vol. III., Diptera, By
FRANCIS WALKER, F.L.S.   8vo, with 10 Plates, 25s.

ANTIQUARIAN.

Bewick's Woodcuts. Impressions of Upwards
2000 Woodblocks, engraved, for the most part, by THOMAS and JOHN BEWICK; including Illustrations of various kinds for Books, Pamphlets, and Broadsides; Cuts for Private Gentlemen, Public Companies, Clubs, &c.; Exhibitions, Races, Newspapers, Shop Cards, Invoice Heads, Bar Bills, &c. With an Introduction, a Descriptive Catalogue of the Blocks, and a List of the Books and Pamphlets illustrated. By the Rev. T. HUGO, M.A., F.R.S.L., F.S.A. In one large volume, imperial 4to, gilt top, with full-length steel Portrait of Thomas Bewick. £6 6s.

The Bewick Collector and Supplement. A
Descriptive Catalogue of the Works of THOMAS and JOHN BEWICK, including Cuts, in various states, for Books and Pamphlets, Private Gentlemen, Public Companies, Exhibitions, Races, Newspapers, Shop Cards, Invoice Heads, Bar Bills, Coal Certificates, Broadsides, and other miscellaneous purposes, and Wood Blocks. With an Appendix of Portraits, Autographs, Works of Pupils, &c. 292 Cuts from Bewick's own Blocks. By the Rev. THOMAS HUGO, M.A., F.S.A. 2 vols. demy 8vo, price 42s.; imperial 8vo (limited to 100 copies), with a fine Steel Engraving of Thomas Bewick, £4 4s. The SUPPLEMENT, with 180 Cuts, may be had separately; price, small paper, 21s.; large paper, 42s.

Sacred Archæology; a Popular Dictionary of
Ecclesiastical Art and Institutions, from Primitive to Modern Times. Comprising Architecture, Music, Vestments, Furniture Arrangement, Offices, Customs, Ritual Symbolism, Ceremonial Traditions, Religious Orders, &c., of the Church Catholic in all Ages. By MACKENZIE E. C. WALCOTT, B.D. Oxon., F.S.A., Præcentor and Prebendary of Chichester Cathedral. Demy 8vo, 18s.

A Manual of British Archæology. By CHARLES BOUTELL, M.A. Royal 16mo, 20 Coloured Plates, 10s. 6d.

Shakespeare's Sonnets, Facsimile, by Photo-Zincography, of the First Printed Edition of 1609. From the Copy in the Library of Bridgewater House, by permission of the Right Hon. the Earl of Ellesmere. 10s. 6d.

Man's Age in the World according to Holy Scripture and Science. By An ESSEX RECTOR. Demy 8vo, 8s. 6d.

The Antiquity of Man; an Examination of Sir Charles Lyell's recent Work. By S. R. PATTISON, F.G.S. Second Edition. 8vo, 1s.

MISCELLANEOUS.

On Intelligence. By H. TAINE, D.C.L. Oxon. Translated from the French by T. D. HAYE, and revised, with additions, by the Author. Part I. 8s. 6d. Part II. 10s., or, complete in One Volume, 18s.

The Birds of Sherwood Forest; with Observations on their Nesting, Habits, and Migrations. By W. J. STERLAND. Crown 8vo, 4 Plates. 7s. 6d. coloured.

The Naturalist in Norway; or, Notes on the Wild Animals, Birds, Fishes, and Plants of that Country, with some account of the principal Salmon Rivers. By the Rev. J. BOWDEN, LL.D. Crown 8vo, 8 Coloured Plates. 10s. 6d.

The Zoology of the Voyage of H.M.S. *Samarang*, under the command of Captain Sir Edward Belcher, C.B., during the Years 1843-46. By Professor OWEN, Dr. J. E. GRAY, Sir J. RICHARDSON, A. ADAMS, L. REEVE, and A. WHITE. Edited by ARTHUR ADAMS, F.L.S. Royal 4to, 55 Plates, mostly coloured, £3 10s.

Travels on the Amazon and Rio Negro; with an Account of the Native Tribes, and Observations on the Climate, Geology, and Natural History of the Amazon Valley. By ALFRED R. WALLACE. Demy 8vo, with Map and Tinted Frontispiece, 18s.

A Survey of the Early Geography of Western Europe, as connected with the First Inhabitants of Britain, their Origin, Language, Religious Rites, and Edifices. By HENRY LAWES LONG, Esq. 8vo, 6s.

The Geologist. A Magazine of Geology, Palæontology, and Mineralogy. Illustrated with highly-finished Wood Engravings. Edited by S. J. MACKIE, F.G.S., F.S.A. Vols. V. and VI., each, with numerous Wood Engravings, 18s. Vol. VII. 9s.

The Stereoscopic Magazine. A Gallery for the Stereoscope of Landscape Scenery, Architecture, Antiquities, Natural History, Rustic Character, &c. With Descriptions. 5 vols., each complete in itself and containing 50 Stereographs, £2 2s.

Everybody's Weather-Guide. The Use of Meteorological Instruments clearly Explained, with Directions for Securing at any time a probable Prognostic of the Weather. By A. STEINMETZ, Esq., Author of "Sunshine and Showers," &c. 1s.

Sunshine and Showers: their Influences
throughout Creation. A Compendium of Popular Meterology.
By ANDREW STEINMETZ, Esq. Crown 8vo, Wood Engravings,
7s. 6d.

The Reasoning Power in Animals. By the
Rev. J. S. WATSON, M.A. Crown 8vo, 9s.

Manual of Chemical Analysis, Qualitative and
Quantitative ; for the Use of Students. By Dr. HENRY M.
NOAD, F.R.S. New Edition. Crown 8vo. 109 Wood-Engravings,
16s. Or, separately, Part I., 'QUALITATIVE,' New Edition,
new Notation, 6s. ; Part II., 'QUANTITATIVE,' 10s. 6d.

Phosphorescence; or, the Emission of Light
by Minerals, Plants, and Animals. By Dr. T. L. PHIPSON,
F.C.S. Small 8vo, 30 Wood Engravings and Coloured Frontispiece, 5s.

Meteors, Aerolites, and Falling Stars. By
Dr. T. L. PHIPSON, F.C.S. Crown 8vo, 25 Woodcuts and
Lithographic Frontispiece, 6s.

The Artificial Production of Fish. By PIS-
CARIUS. Third Edition. 1s.

Live Coals; or, Faces from the Fire. By L.
M. BUDGEN, "Acheta," Author of 'Episodes of Insect Life,'
etc. Dedicated, by Special Permission, to H.R.H. Field-Mar-
shal the Duke of Cambridge. Royal 4to, 35 Original Sketches
printed in colours, 42s.

Caliphs and Sultans; being Tales omitted in
the ordinary English Version of "The Arabian Nights' Enter-
tainments," freely rewritten and rearranged. By S. HANLEY,
F.L.S. 6s.

SERIALS.

The Natural History of Plants. By Professor BAILLON, with numerous Wood Engravings. Monthly, 2s. 6d.

The Botanical Magazine. Figures and Descriptions of New and Rare Plants of interest to the Botanical Student, and suitable for the Garden, Stove, or Greenhouse. By Dr. J. D. HOOKER, F.R.S. Monthly, with 6 Coloured Plates, 3s. 6d. Annual Subscription, post free, 42s.

The Floral Magazine. Figures and Descriptions of Select New Flowers for the Garden, Stove, or Conservatory. By the Rev. H. H. DOMBRAIN. New Series, enlarged to Royal 4to. Monthly, with 4 Coloured Plates, 3s. 6d. Annual Subscription, post free, 42s.

Conchologia Iconica. By LOVELL REEVE, F.L.S., in Double Parts, with 16 Coloured Plates, 20s.

Conchologia Indica. The Land and Freshwater Shells of British India. In Parts, with 20 Coloured Plates, 20s.

A Monograph of Odontoglossum. By JAMES BATEMAN, F.R.S. Imperial folio, 5 Coloured Plates, 21s.

Select Orchidaceous Plants. By ROBERT WARNER. 3 Coloured Plates, 10s. 6d.

FORTHCOMING WORKS.

The Natural History of a Flowering Plant. By Prof. DYER.

The Young Collector's Handy Book of Recreative Science. By the Rev. H. P. DUNSTER.

Flora of India. By Dr. HOOKER and Dr. THOMSON.

Natural History of Plants. By Prof. BAILLON. Vol. II.

LONDON:
L. REEVE & CO., 5, HENRIETTA ST., COVENT GARDEN.

www.ingramcontent.com/pod-product-compliance
Lightning Source LLC
Chambersburg PA
CBHW031940230426
43672CB00010B/1990